AF567136

APPLICATIONS OF DECISION TABLES—A READER

Applications of Decision Tables

A Reader

edited by Herman McDaniel
U.S. Civil Service Commission

Brandon/Systems Press, Inc.
Princeton New York London

Library of Congress Catalog Card Number: 71-100998
Standard Book Number: 87769-022-7

First Printing

Printed in the United States of America

Contents

Acknowledgments

Acknowledgment of permission to use materials in this book is as follows:

Chapter 1, by D. T. Schmidt and T. F. Kavanagh, originally appeared under the title, "Using Decision Structure Tables" in *Datamation.* Reprinted with permission from *Datamation,* February, 1964, published and copyrighted 1964 by F. D. Thompson Publications, Inc., 35 Mason Street, Greenwich, Connecticut, 06830.

Chapter 2, by D. T. Schmidt and T. F. Kavanagh, originally appeared under the title, "Using Decision Structure Tables," in *Datamation.* Reprinted with permission from *Datamation,* March, 1964, published and copyrighted 1964 by F. D. Thompson Publications, Inc., 35 Mason Street, Greenwich, Connecticut, 06830.

Chapter 3, by Burton Grad, appeared in *Data Processing, Vol. VIII* (the Proceedings of the 1965 International Data Processing Conference). It is reprinted here by permission of the Data Processing Management Association, Park Ridge, Illinois.

Chapter 4, by T. F. Kavanagh and M. Allen, originally appeared under the title, "The Use of Decision Tables," in *Data Processing, Vol. VI* (the Proceedings of the 1963 International Data Processing Conference). It is reprinted here by permission of the Data Processing Management Association, Park Ridge, Illinois.

Chapter 5 was first printed in the *Seminar on Decision Tables; Two Years of Experience at the U.S. Bureau of the Census,* Washington, D.C., September 30, 1964.

Chapter 6 was originally given as an address by Richard A. Hornseth at the *Seminar on Decision Tables; Two Years of Experience at the U.S. Bureau of the Census,* Washington, D.C., September 30, 1964.

Chapter 7, by H. S. Woodgate, is reprinted with permission from *Datamation,* January 1968, published and copyrighted 1968 by F. D. Thompson Publications, Inc., 35 Mason Street, Greenwich, Connecticut, 06830.

Chapter 9, by HERBERT R. LUDWIG, is reprinted with permission from the *Journal of Data Management,* January 1968, copyrighted by the Data Processing Management Association, Park Ridge, Illinois.

Chapter 10, by ROBERT C. FIFE, was published under the title "Decision Tables" as part of the *Proceedings of the Univac Users Association,* Spring, 1966. It is used here by permission.

Chapter 11, by L. E. LESKINEN, first appeared in the *Proceedings of the Univac Users Association* under the title "General Purpose Systems Design and Programming Techniques," Fall, 1966. It is reprinted here with permission.

Chapter 12, by THOMAS B. GLANS, first appeared under the title, "Progress in Decision Table Applications," in the Systems and Procedures Association's publication, *Ideas for Management—1963.* It is used here with permission.

Chapter 14 is adapted with permission from *EDP Analyzer,* May 1966, published by Canning Publications, Inc., 134 So. Escondido Ave., Vista, California.

Chapter 15, by DONALD DEVINE, appeared in *Data Processing, Vol. VIII* (the Proceedings of the 1965 International Data Processing Conference). It is reprinted here with permission of the Data Processing Management Association, Park Ridge, Illinois.

Chapter 16, by H. I. MEYER, appeared in *Data Processing, Vol. VIII* (the Proceedings of the 1965 International Data Processing Conference). It is reprinted here with permission of the Data Processing Management Association, Park Ridge, Illinois.

Chapter 17, by L. E. LESKINEN, originally appeared under the title, "Programming in a Modular Form With or Without Decision Tables," *Proceedings of the Univac Users Association,* Fall, 1966. It is used here with permission.

Chapter 18, by SOLOMON L. POLLACK, is from the Systems and Procedures Association publication, *Ideas for Management*—1966. It is used here with permission.

APPLICATIONS OF DECISION TABLES—A READER

1

The Use of Decision Tables

In today's industrial world, systems are being developed to convert customer orders into finished products automatically, so they say. In designing and implementing such complex systems, techniques are needed to reduce applications effort. This means the work required to understand and define the problem, develop and program a solution, and also provide the often neglected documentation.

Decision structure tables meet these requirements. They provide a simple method for recording logic, so that all elements of a decision are precisely defined. Tables make it possible for managers, engineers, accountants, etc., to use computers directly. They eliminate much subsequent programming and coding effort. And the direct use of decision structure tables in a computer opens new horizons for computer applications.

But with any relatively new technique, there are fundamental questions that must be answered:

1. Has anyone "made money" using decision structure tables?
2. How are problems "structured"?
3. Specifically, where are decision structure tables being used in working programs?
4. If decision structure tables are good, why isn't everybody using them?

This article attempts to answer these questions and give practical information on the actual use of decision structure tables.

Interest by Profession

Decision structure tables have been the subject of many articles and presentations at professional society meetings, such as the CODASYL-JUG Symposium of 1962 that introduced DETAB-X. Of the 500 persons who attended, nearly 250 remained for a follow-up tutorial session. CODASYL's Systems Group has been especially active in promoting interest in structure tables.

Computer manufacturers have also shown interest by fostering the development of programs that would accept decision structure tables for compiling and processing. General Electric, for example, introduced TABSOL (*TAB*ular *S*ystem *O*riented *L*anguage) as an integral part of GECOM—its generalized, automatic compiler for the GE 215, 225, and 235. IBM and RAND cooperated in the development of FORTAB, a structure table preprocessor to FORTRAN.

Where and How Decision Structure Tables Are Used

Decision structure tables are best applied where there are numerous detailed, interacting decisions involved in a problem solution. The same problem flow charted would have many levels. It would be quite complex and contain "Merry Christmas trees" with many interlocking branches.

For example, in the machine shop a lathe operator must be provided with as many as 30 to 50 values before he can start his job. He must be supplied with tool description, speed, feed, depth of cut, operation sequence, etc. Each of these items is dependent on many factors, such as the machine, the material, and the specific operation. In the office, production and inventory control specialists must know current demand, trends, cost, shelf life, and many other values in order to develop intelligent replenishment orders. Similar illustrations could be developed for almost any situation in any business.

These are the kinds of intricate decisions where structure tables have proven especially valuable. Often the complexity and extensiveness of these detailed decisions are drastically underestimated—much to the chagrin of overly enthusiastic computer applications engineers, their customers, and their managers. But approached with proper respect, these day-by-day decisions offer one of the brightest potentials for computer applications work.

Writing Decision Structure Tables

Initially, decision structure tables require some persistence until the skill is mastered. But this is true of any new technique that breaks with established tradition. Actually, writing structure tables is quite simple. It's the process of logical thought which creates the difficulty. Structure tables merely highlight the stumbling block. For this reason, some ground rules were devised to help write tables more easily.

In these ground rules there is a shift in emphasis from flow charting. Flow charts emphasize activities, sequence, and flow. In working with flow charts, the first step is usually to identify the major activities, and then to identify the sequence in which they would be performed. These hi-level charts are too macroscopic. Further refinement produces a succession of lower level, more detailed, and more voluminous flow charts. The decision logic that controls the activities isn't really incorporated until you get down several levels into detail.

When working at more detailed levels, the programmer must determine how any given decision should be made. The answer probably depends on the value of many different parameters, each of which in turn may be the result of preceding levels or branches. But at this stage most flow charts consist of many balloons, arrows, boxes, lines, etc. in the programmer's unique notation. Flow charts, particularly at the computer code leve, contain many yes-no decisions and triple choices —minus, zero, plus. This is a rather small selection. The net effect: Flow charts must be complex to adequately define and solve real world problems.

The shift in emphasis, therefore, is from a flow of activities—which considers decisions secondary—to a technique that considers the decision logic primary.

Six Ground Rules

From the beginning, we have always faced the problem of describing to others how a decision problem or system is "structured." Over a period of time, a six-step procedure evolved. These six ground rules provide a step-by-step method for writing structure tables. Each step focuses attention on an important element in the decision-making process. We recognize there is much in these rules that is not profound. But they do describe how a decision is "structured." Further, they provide a consistent approach which has been used successfully to instruct beginners. It is not suggested that these rules must be followed rigorously after proficiency has been gained—but even so, they are a good discipline.

Try writing your decision structure tables using these six steps. They permit you to concentrate on specific decisions—one at a time. See if this approach doesn't help clarify how each decision relates to the total decision-making problem.

It is not unusual to repeat each step several times. Each refinement enables the table writer to improve his decision logic. Define Step 1 as best you can, and then try Step 2. In Step 2 you may uncover something which could correct, broaden, enlighten, or make Step 1 more complete. This isn't an error, it's to be expected. This iteration removes logical errors and inconsistencies; simplifies solution logic; makes it more general, and improves the quality of the answers.

Each of the six steps is described below. To demonstrate how the rule is applied, an actual application taken from General Electric's X-Ray Department is outlined. The case history shows how manufacturing engineering specialists used decision structure tables to develop the logic for detailed operating instructions required to make turned parts.

RULE 1—DEFINE SPECIFIC BOUNDARIES FOR THE PROBLEM

Define the objective in meaningful terms. You can't solve all the world's problems, nor can you structure all business decision-making. These objectives are too broad, too general. Like any other computer application, structuring projects must have a specific purpose. Hopefully, the resulting work will have economic value. Perhaps it will solve a decision-making problem, simulate a system, or otherwise be useful. The objective should contain the basis for performance specification and measurement. In the illustration, for example, X-ray must tell factory operators which parts to make and how to make them. Perhaps the computer can do it more uniformly and more accurately. Perhaps the present paperwork system takes too long. Perhaps the only objective is to reduce cost through mechanization. But whatever it is, the ultimate objective must be specified in sufficient detail to provide working goals for the project. These statements will assist in establishing project schedules and manpower and computer requirements.

Set limits and ranges. Because there are as many ways of making turned parts as there are turned parts themselves, it is necessary to detail the objective much further. First, what is the variety of turning equipment in the shop? (Of course, if new equipment is to be purchased, the picture changes.) Then too, what about boring mills and other pieces of equipment currently available that could be used to "turn parts" under extreme situations? True, they *could* be used but it's better shop practice to use turret lathes. The decision to include or exclude off-standard machines is an important one delimiting the scope of the project.

Additional work must be done to identify and define the characteristics of the various parts being studied: Diameter, length, material, finish, threads, configuration, etc. "Diameter," for example, is important because of lathe feeds and chucking characteristics. Such limits effectively define boundaries of the planning system. Suitable ranges must be established for each parameter. For example:

Diameter	0.250 min. . . . . 2.000 max.
Length	0.250 min. . . . . 5.000 max.
Material	copper, brass, steel, etc.

Setting these limits and ranges is an extremely important step because you are essentially setting a ceiling on the ultimate value of your work. Obviously, the broader the boundaries, the more problems the resulting tables will be able to solve—but the difficulty will also rise, as will the number of tables—sometimes exponentially. The task, therefore, is to define the real problem. Anything more may have limited value in application; anything less is unsatisfactory.

For example, if the system doesn't cover enough parts in the shop to make it worthwhile, then the work is of limited use. The other extreme would be to encompass so much that you're back to economics —it's so expensive and takes so long to get results, that it's just not worth it.

If the objective or problem specification cannot be nailed down, there's some doubt as to whether you should proceed on the project —structure tables or not!

Rule 1—Example—Case History: Manufacturing Operation Structure Tables for Turned Parts

A. *Define the objective*, that is, factory operator instructions.

1. Produce manufacturing operation planning for making turned parts on lathes—to avoid individual planning of every part.
2. Planning has to include:
   a. Machine selection
   b. Operation description
   c. Operation sequence
   d. Tool selection
   e. Setup information
   f. Running instructions
   g. Feeds and speeds
   h. Time standards

B. *Set limits and ranges* for turned parts.

1. X-ray further limited study to parts that were:
   a. Made on turret lathes in X-ray's machine shop
   b. Made from continuous bar-stock (this could easily be expanded to include chucked parts if the holding or chucking criteria were included—but they weren't).
2. The planning system must reflect existing shop practices of the department.
3. Define characteristics of parts:
   a. Physical size
   b. Physical shape
      Simple parts—washers, spacers
      Complex parts—shafts
   c. Materials
   d. Surface finish
   e. Surface modifications
      Bevels
      Chamfers, rounds
      Threads
      Tapers
      Knurls
      Grooves

4. Define characteristics and capabilities of machines and tools:
   a. Machine capabilities (size, feeds, speeds, horsepower tolerances)
   b. Tools capabilities (finish, clearance)
5. Define the ranges or allowable values for each characteristic:
   a. 7 external diameters (maximum per part)
   b. 3 internal diameters (maximum per part)
   c. 20 different materials (defined initially)
   d. surface finish between 16 and 250 RMS
   e. tolerance capabilities of the machine and cutting tools in terms of specific materials
   f. and many, many others. . . .

To do this, X-ray made an extensive study of existing blueprints, operation planning records for current parts, and existing machine tools. This enabled them to determine the requirements of the planning system more realistically in terms of actual needs.

RULE 2—ENUMERATE INDIVIDUAL "ELEMENTARY" DECISIONS

Rarely are real world decisions the result of one single stroke of intuition or analysis. More typically, the results, which sometimes appear deceptively simple, are the culmination of very many smaller decisions and evaluations. In some respects, it is these small decisions which present most difficulty. Similarly, real world decisions rarely consist of a single numeric value, or a simple yes–no. Most often, many values and actions must be specified to completely define a decision so it can be implemented. In Step 2 we are trying to break down the objective established in Step 1 into a group of smaller decisions or areas of activity. Indeed we want to get down to the smallest meaningful detail of solution—hence the term "elementary." In brief, what do I have to know to solve the problem or make the decision? For example, in the turning project, these are elementary decisions: What are the speeds and feeds? What coolants or lubricants should be used? What size stock? Which machine?

Note the emphasis on small, relatively simple decisions. Each of these small decisions eventually will serve as the basis for one or more structure tables. If you are already flow chart oriented, macro-block diagrams may help break down the problem into elementary decisions in an organized way.

From this point on, attention will be centered on individual elementary decisions. All remaining steps in the six-step sequence should be performed for *each decision* that was developed in Step 2.

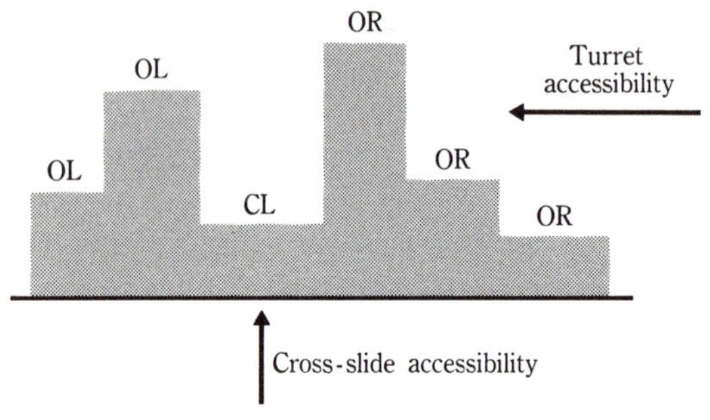

FIG. 1-1

Rule 2—Example

Here are some examples of decisions and problems which must be solved in order to develop operation planning for turned parts:

In turning parts on a lathe, a first consideration is the *L/D* (length over diameter) ratio. This ratio indicates whether the part can be made accurately, that is without bending under tool pressure. If the length is excessive, intermediate supports can be added, or possibly another manufacturing process should be used.

All that has been established now is that the *L/D* ratio must be checked—the value determination and refinement will come later.

Each diameter of the part must be classified as open right (OR), open left (OL), or closed (CL). This decision area—that is, the logic to assign these surface classifications from a part description—is quite complex. It is needed later in the decision making system to determine the accessibility of any surface to tools mounted on the turret (end) or cross-slide (front) of the lathe. (See Fig. 1-1)

Here are some examples of other decision areas—each of which will require many small individual decisions—too numerous to include in this article:

1. Calculate length and depths of cuts
2. Sequence each cut

3. Select machine
4. For internal and external diameters
   a. Select process
   b. Modify surfaces
   c. Assign tools
   d. Sequence tools
5. Determine feeds and speeds
6. Determine time standards
7. Format printout

One simple decision that is required would be the selection of raw bar-stock from which to make a part. We will use this decision as an example to show how each subsequent step further develops the elementary decision.

RULE 3—DEFINE THE NECESSARY OUTCOMES

Rule 2 identified, in general terms, the individual decisions that must be made. These decisions must now be defined in terms of specific outcomes, known as results or actions. These outcomes may be specific parameter values, procedural actions, or output results. Any of these may be an intermediate outcome required later in the total problem solution.

The results of most elementary decisions consist of several parameter values. These different outcomes are required to completely define all variables in the solution. For example, it is not enough to decide that "yes," a coolant must be applied during machining operations. Typically, processing a machined part requires several machining operations: turning, boring, drilling, tapping, etc., and the decision to use, or not to use, a coolant must be made for each individual operation. If the decision to use a coolant is made, then more specifically what coolant? Coolants come in many varieties—such as oils, emulsions, etc. And even further, which oil or emulsion? Additionally, the method of application must also be defined, for instance, flood or mist.

Specifying outcomes is quite different when structuring "existing" systems or practices versus "proposed" or new systems. In existing systems the result parameter and parameter values are generally fairly well defined by the current situation. For this reason tables may be purposely written to produce these results. For example—tool selection decisions may be limited to tools that exist in the shop at the present time. This may not result in optimum tool selection, but hopefully they are the best existing tools for the job.

If the system was a new system—and no tools existed—then the structuring approach could provide an analysis technique for determining what tools would be required for "optimum" performance. Then the planning system could be designed around these tools. Quite possibly, over a period of time, tool requirements would change, and the structure tables would have to be updated.

In addition to parameter names, the decision results, or outcomes, can also signal activities to be performed or indicate the elementary decision to be considered next. PERFORM is an action word which tells the structure table to execute some specific logic or activity designated in the column below. Typically the activity is a common one, such as searching a list, writing a report, or making some common calculation —perhaps taking a square root. The important point about PERFORM is that when the activity has been completed, control returns to the structure table from which the PERFORM was originally issued.

GO TO is analogous to the command GO. This transfers control unconditionally—and without restraint—to some other structure table. This is the mechanism used for moving from one structure table to another; it is the linkage between tables. It essentially replaces the arrow lines on flow charts.

Approaching the problem more classically, the result functions are concerned with the degrees of freedom to be allowed in the decision system. For reasons of cost and complexity, these result parameters must be established with care. Finally, any single decision may result in multiple outcomes. Typically, in industrial applications, one decision will result in many actions. For this reason, the action portion of a structure table is frequently larger than the condition side.

Rule 3—Example

In Rule 2 one elementary decision was selected for further explanation: "What bar stock should be used to make this part?"

Defining the necessary outcomes requires the table writer to think about how bar stock is described. What does someone have to know to differentiate one kind of bar stock from another? It turns out that bar stock is described by specifying:

Material: Brass, steel, etc. Sometimes various coded designations of chemical content are required to further identify material—AISI 1040 tool steel, for example.

Shape: Round, hex, rectangular, etc. This is the shape of a cross-sectional area. Sometimes it is also necessary to state whether the material comes in bars, strips, coils, etc. and tell what the lengths are.

Size: Associated with a shape is a commonly accepted size designation. Round bars are specified by diameter; hexagonal shapes by the distance between opposite flats; but rectangular shapes require two dimensions.

Finish: Rough, cold rolled, ground, etc. This is particularly important in developing planning for the first machining operation.

In addition to this basic description, additional information might be required for more sophisticated planning systems which automatically calculate best speeds and feeds. For these calculations, material hardness, ductility, cost, etc. are usually required.

In X-ray's project, the manufacturing planners always used bar stock. Hence, it was unnecessary to write tables covering coil stock—at least for the present. Of course, if the Department later decides to install some new equipment to handle coil stock, then the structure tables will have to be updated. Ease of maintenance is as important for manufacturing information systems as it is for factory machines.

Similarly, X-ray usually worked with cold rolled finishes, hence this parameter did not vary.

To simplify this illustration further, material specification—a true variable in X-ray—has been arbitrarily dropped. Obviously, X-ray's tables actually contain a full complement of different materials.

The remaining result parameters—bar-stock size and shape—require a decision. This decision will be a selection from the available alternative values developed in Rule 4 below. The result parameters can now be entered in the "result or action headings" as follows:

	Bar-stock Shape	Bar-stock Size	

In another interesting example, engineers involved in wound coil design once decided to introduce some "coil standardization." Two major variables were wire diameter and number of wire turns per coil. These two variables determine general coil properties. The engineers "standardized" the number of wire turns per coil, resulting in a long list of wire diameters. Stocking many different wire sizes creates some inventory control problems for the factory.

Manufacturing, on the other hand, produces the coils by winding them on an arbor. The number of turns in the coil can easily be controlled by the number of arbor revolutions. Using decision structure tables uncovered this incompatibility—resulting in a coil redesign with complete flexibility in the number of turns and a substantal reduction in the "standard" wire sizes. Now, everyone was happy.

RULE 4—DEVELOP VALUE STATES FOR EACH ALLOWABLE OUTCOME

Each result parameter will usually have a variety of values within the limits and ranges of the basic problem outlined in Rule 1. Considering each and every possible result state could make tables impossibly large. For example, suppose there are five allowable values for each of two parameters. If all combinations were allowed, then 25 different result values would be possible. Twenty-five result values are usually, but not always, unnecessary. Frequently, it is possible to develop simple formulas or equations to describe how a result value changes as a function of various decision parameters. Over-specification of outcome values may also create unnecessary results.

The approach to result values, as well as result parameters, differs greatly with "existing" versus "proposed" systems. Values for existing systems must reflect current usage and practice. In proposed systems we have the opportunity, and obligation, to include the best values for the decision. This, however, is not done easily. It usually requires considerable study and analysis to determine these "optimum" values.

Results may be other than specific values assigned to outcome parameters (data results). The values may be the names of "procedural" or "logic" results. They consist of routines or tables to be solved next as indicated by a GO TO or PERFORM result parameter. They could also be "input/output" results associated with READ or WRITE.

Rule 4—Example

Having defined the bar stock size and shape as necessary outcomes, it is now desirable in the case of an existing system to establish the present value states for these parameters. For instance, there may be only round, square, and hexagonal bar stock shapes in the stock room. The round shape may exist in ¼, ½, ¾, 1, 1½, and 3 inch diameters; the square in ½, 1, and 2 inch sizes, and the hexagonal in ¼, ⅜, ¾, 1, 1½, 2, and 3 inch sizes. The raw material stocking program is often outside the boundary of a single manufacturing planning study, hence the available materials, sizes, and shapes are "fixed." They must be used the best way possible.

Conversely, it could have been decided, because of the relative ease of obtaining different bar stock shapes and sizes to specify only those sizes and shapes which minimize scrap. In reality, however, manufacturing planning and inventory control should analyze bar stock requirement to balance volume purchase prices against the costs associated with buying over-size stock and removing excess materials. This analysis should produce a list of "optimum" bar stock shapes and sizes.

From a computations viewpoint, a geometric stock plan might make it possible to calculate the bar stock using a formula, thereby reducing the size of the table considerably. However, this computational convenience is of secondary importance. Primarily, the tables must record reality. Actual bar stock shapes and sizes are posted in the table following.

Other decision situations in the X-ray study, such as machine tool selection, could not be changed so easily—thus the planning was tailored to the available equipment.

RULE 5—DEVELOP DECISION PARAMETERS AFFECTING EACH DECISION

Decision parameters are qualitative factors affecting a decision. So far all possible decision results and outcomes have been specified, but little has been said about the selection between these alternatives.

The first step in selecting an alternative is to establish the parameters which control the choice. Decision parameters can usually be identified by asking "On what does the decision depend?" At this time, identification of the decision parameters is enough—the assignment of values is done later in Rule 6.

Bar-stock Size	Bar-stock Diameter
Round	0.2500
"	0.5000
"	0.7500
"	1.0000
"	1.5000
"	3.0000
Square	0.5000
"	1.0000
"	2.0000
Hex	0.2500
"	0.3750
"	0.7500
"	1.0000
"	1.5000
"	2.0000
"	3.0000

Since the decisions are "elementary," the outcomes are usually a function of only two or three variables. Seldom does a single elementary decision require more than five parameters. In fact if it appears that five or more variables are required, it is doubtful the decision is truly elementary. Apparently life's like that. Decision parameters should uncover true "cause and effect" relationships which really describe why various different alternatives are selected. This step can be a very revealing process in itself.

To correctly establish the decision parameters will require a thorough understanding of the decision. Without this understanding, it is doubtful proper structuring can be done. But with it, many of the previous goals, understanding, documentation, reduced costs, and computer applications can be greatly facilitated.

The structuring approach has some built-in indicators when improper or illogical relationships are likely to occur. An excessive number of decision parameters is one extreme; extreme difficulty in establishing relationships between decision and result parameters is another; decision parameters whose values are arbitrary, such as non-significant identification numbers rather than physical entities such as dimensions, colors, weights, voltages, current, etc. is still another. On the other hand, if the situation seems extremely clear, and the decisions are relatively small, odds are that the approach is probably sound.

With this understanding of the type of intimate knowledge required for structuring, it is obvious that those best qualified to do the work are the managers, engineers, accountants, and other functional specialists actually responsible for it.

Rule 5—Example

In the X-ray example, the material shape and size were desired outcomes. Accordingly available values were established for each. The problem now is to establish the rules for selecting any of these specific alternatives. This example is an extremely simple case. Most others will probably be somewhat more difficult.

Generally, the large diameter of a part will govern the size of the raw material. Similarly, the shape of the largest diameter probably will determine the shape of the raw material, particularly if the stock configuration (shape and size) is used "as is" for the largest diameter. This might be appropriate for such parts as bolts, studs, spacers, etc. If the part being planned was cylindrical then the problem is reduced to one of selecting the smallest diameter round stock from which the part can be made. Sometimes it will be necessary to select the next largest diameter because of stock unavailability or because the largest diameter is to be machined.

The actual number of decision parameters for a turned part can be considerable—particularly for complex parts. Values for these decision parameters in any one problem must typically be supplied from problem input. Thus it is important to develop as much as possible from each input parameter. X-ray inputs generally consisted of engineering dimensions, tolerances, surface finishes, thread sizes, etc., all basic physical characteristics of turned parts. The important point is that when decision parameters are being developed in Rule 5, the table writer is also specifying "input" parameters. In a very real sense he is automatically developing his input data requirements.

One might also visualize another structure table system to design turned parts. The output from such engineering design systems could provide input for manufacturing planning systems automatically. This is not an unrealistic proposition.

The decision and input parameters pertinent in our illustration are:

Largest OD	Largest OD Shape	Bar-stock Shape	Bar-stock Size
		Round	0.2500
		"	0.5000
		"	0.7500
		"	1.0000
		"	1.5000
		"	3.0000
		Square	0.5000
		"	1.0000
		"	2.0000
		Hexagonal	0.2500
		"	0.3750
		"	0.7500
		"	1.0000
		"	1.5000
		"	2.0000
		"	3.0000

RULE 6—DEVELOP DECISION PARAMETER TESTS AND VALUES

The next and last step is to assign specific decision parameter tests and values. These tests are used to identify the exact breaking point between alternative results or outcomes. Recall, from the insert describing structure table fundamentals, that the decision side of the structure table functions as a pictorial IF–THEN condition statement which relates decision parameter values directly to specific result values. There will be a set of entries for every outcome. If there isn't, there may be trouble. Perhaps the result is not legitimate; or maybe there are missing decision parameters. If it's the last row of the table, the outcome may be a convenient catch-all which is selected whenever the special conditions tested in earlier rows do not apply. Here, the good work done in Step 1 defining the problem in great detail will become very helpful. Many of these numbers will now take their place in structure tables.

When it is not possible to assign specific numbers or ranges to decision parameter value states, use ordered sets of terms—for example good, better, best. The fact that a value state is not a number shouldn't cause difficulty. Actually, the only criterion is that the value state be descriptive of the decision parameter.

The comments made earlier about avoiding exhaustive enumeration of possible result values are equally appropriate on the decision side of the table. Wherever possible test minimums and maximums. When one or more values of a decision parameter satisfy a test, string them together in an OR statement. Similarly, arithmetic expressions can also be used as decision parameter values.

Filling out decision parameter tests and values is similar to designing a screen to filter a given problem into the correct solution row. All the normal COBOL relational operators are available for formulating tests: LS, NLS, GR, NGR, EQ, NEQ. But overrefinement can be costly. The decision side of the structure table should not be more discriminating than the results. If several rows of tests and values yield the same results, the decision parameter value states may be too precise. Such overspecification increases both the size and the solution time of the tables. Also, it implies a degree of precision which is not present in the system.

Rule 6—Example

In the example, the relationship of largest OD shape to bar stock shape is rather simple, particularly in the round state. It becomes a little more complex when square and hex stock are turned down to a round shape.

When turning square and hex shape parts, the largest "OD" must correspond exactly with the stock size.

In this table round stock is not available in 3/8 and 2 inch sizes—but hex stock is. A decision therefore was made to turn down hex stock to the round size for certain OD's. Thus, for two sizes—even though the largest OD is to be round—hex stock will be used!

With the six rules outlined, two key points should be emphasized:

• First, going through the six-step procedure once doesn't mean the job is complete. Going through several times shows how the structure tables interact with one another. The shrewd table-writer goes back to Rule 1 again and again to rethink the problem and see if he has done a thorough job. By the second time around, he will see ways in which the logic might be simplified, corrected, and made more complete. Repeating the process is not an error. It is a way to achieve more significant results. Remember, it is these same tables that can be used to "solve" problems manually or can be compiled into a computer program for mechanized problem solution. Further, they document the logic for

Largest OD Shape	Largest OD Size	Largest OD Size	Bar-stock Shape	Bar-stock Size
Round	GR 0	NGR 0.230	Round	0.250
"	GR 0.230	NGR 0.355	Hex	0.375
"	GR 0.355	NGR 0.480	Round	0.500
"	GR 0.480	NGR 0.730	"	0.750
"	GR 0.730	NGR 0.980	"	1.000
"	GR 0.980	NGR 1.475	"	1.500
"	GR 1.475	NGR 1.900	Hex	2.000
"	GR 1.900	GR 2.950	Round	3.000
Square	GR 0	NGR 0.490	Square	0.500
"	GR 0.490	NGR 0.990	"	1.000
"	GR 0.990	NGR 1.975	"	2.000
Hexagonal	GR 0	NGR 0.230	Hex	0.250
"	GR 0.230	NGR 0.350	"	0.375
"	GR 0.350	NGR 0.725	"	0.750
"	GR 0.725	NGR 0.975	"	1.000
"	GR 0.975	NGR 1.470	"	1.500
"	GR 1.470	NGR 1.965	"	2.000
"	GR 1.965	NGR 2.960	"	3.000

educating newcomers and communicating with others. In brief, they're worth the effort.

• Second, if at all possible, focus on the general engineering solution to the basic problem "general" because the tables should provide answers to a whole class of related or similar problems; "engineering" because you are looking for efficient, reliable, economic solutions, but not necessarily optimal solutions.

In developing decision structure table values and parameters it is usually necessary to consider specific cases, as much of our present approach to documentation records the answers to specific problem situations. However, when writing structure tables, it is important to step back from specific solutions and try developing the logic for as large a set of conditions and circumstances as possible. For example, engineering blueprints and tabulated drawings most often record the results of a designer's solution to a variety of past problems each solved individually. In writing structure tables we are less concerned with identifying these past solutions than we are with uncovering the decision logic which generated them. If we can successfully uncover the logic for making these decisions, we may be able to reapply it to other sets of circumstances when they arise.

The real power of structure tables is not so much in a table look-up or file reference system as it is in uncovering the decision procedure. This increases the power of the logic so that it can handle decision situations which have never occurred previously but do lie within the bounds of the decision system.

This article, by D. T. SCHMIDT and T. F. KAVANAGH, originally appeared under the title, "Using Decision Structure Tables" in *Datamation.* Reprinted with permission from *Datamation,* February, 1964, published and copyrighted 1964 by F. D. Thompson Publications, Inc., 35 Mason Street, Greenwich, Conn. 06830.

2

Manufacturing Applications of Decision Tables

This article emphasizes manufacturing applications because most of our experience is in this area. Decison structure tables coupled wth computers are paying off because they:

- define and think through manufacturing problems, often providing new insights and understanding which have led to improved performance.
- formulate and record decision systems for subsequent use and communication.
- simplify computer implementation where mechanization is desirable.
- get manufacturing to use computers.

There is a wealth of potential computer application in manufacturing. They offer great opportunity. Without structure tables, application costs would be exorbitantly high. It is easy to learn how to use decision structure tables, and, further, the user requires minimum computer knowledge and background. Later in this article a structure table application using computers is described—PRONTO.

Benefits of Decision Structure Tables

Experience in General Electric shows decision structure tables can reduce actual time spent in problem definition and coding. Users report up to 90% lower programming costs because detailed flow charting and coding are virtually eliminated. Other users report 50% lower application costs because of this, plus improved systems design and check-out efficiency.

Others have been attracted to structure tables because conventional flow charting and coding techniques proved unrealistic for their problems. Also, structure tables are easy to maintain and when used in conjunction with a computer, they offer a level of accuracy unequalled in manual sytsems.

Another benefit that recently came to the forefront is more efficient documentation. Engineers have long recognized the need for good documentation; for decades draftsmen have laboriously made detailed drawings of every part, assembly, and product. Dimensions, tolerances, materials, finishes are all carefully noted. Frequently, each drawing is given an identification number signed by the draftsman and approved by a supervisor. Precise records are made of all changes—who made them, when they take effect, whether the modification is interchangeable, etc. This defining of every part in a product—every nut, screw, and washer—is important. But is it less important than the decision logic used to manufacture the same parts, to pay the workers, or to bill the customer who purchases them?

In a very real sense, decision structure tables provide a natural "blueprint" for decision. In this simple analogy, the part itself becomes the elementary decision. A structure table is like a blueprint. Structure tables can delineate groups of decisions, as assembly drawings and outline drawings show groups of parts. Decision logic can be just as vital to a business as its engineering drawings. It should be just as easy to find out how detailed decisions are made as it is to locate the design of a part.

Structure tables also provide an opportunity to improve specialist–manager communication in both directions. On one hand, the specialist receives a much more precise definition of requirements from his manager—while the manager, in turn, receives feedback from the structure tables to assure himself that the detailed decision-making system implements his management objectives. The command and control aspects of decision structure tables offer great promise.

Some have asked if it isn't just as easy to forget a pertinent condition when making up a structure table as it is to overlook a test in flow charting. Quite the contrary—the rigid, uniform tabular format makes it more difficult to make mistakes. Each table entry requires individual attention. The tables also contain some built-in reasonableness checks. The extensive preliminary specification of objectives and the determination of elementary decisions provide overall structure to the problem. The sequential, one at a time, analysis of each elementary decision, developing decision parameters, their related tests and values, and results or outcomes with their values are all designed to minimize the chance for error.

Others have asked how useful decision structure tables will be to the informed senior programmer. Certainly one of the problems faced by senior programmers—all programmers—is applications that come with inconsistent and incomplete problem definition. Considerable time and effort must be spent trying to uncover and resolve these difficulties so that complete, consistent specifications can be written. But if the application came to the programmer defined with sets of decision structure tables in the first place, there would be substantially less emotion in computer applications work. Also, the informed senior programmer would have to spend much less time learning all there is to know about his prospective client's problem.

If the senior programmer is relieved from handling this applications work, he can then concentrate on making existing applications more computationally efficient. There are also integrated operator systems to be designed and programmed to make it easier for the computer to handle all these different problems. The programmer might even build a structure table processor. In fact, he might find tables very useful in structuring the logic of these applications.

Right now, many informed senior programmers are working as "general applications practitioners" when actually they should be working as computer specialists.

After the application has floundered and no one knows what's wrong, then the informed senior programmer must come in and remedy the situation. It would be a lot easier for him to handle the situation if decision structure tables were used.

The application described below illustrates these benefits. PRONTO is a program to automatically convert conventional, manual manufacturing planning to the considerably more elaborate and complex instructions required to operate numerically programmed machine tools.

Development of "PRONTO"

The introduction of automatic electro-mechanical machine tools created demands for a new kind of planning data. Very detailed coded information was necessary to position a part in relation to a tool. Depth control and tool selection were also necessary. Tool motion patterns for operations such as tapping, drilling, and milling were required. Proper feeds and speeds had to be specified, and provision had to be made for auxiliary functions such as turning coolants on and off.

Instructions required for numerically controlled (NC) machine tools were far more precise. They varied greatly from the routine planning instuctions used with conventionally-operated equipment. This preciseness was necessary because the NC machine tool replaced the control and judgment of the human operator.

Using the NC system presented formidable problems. Not only was a high degree of sophisticated planning needed, but a new and unfamiliar "language" was required. Also, standards did not exist and each machine tool required unique inputs. The number of routine, repetitive calculations was enormous and sometimes very complex; this meant the error potential was high and these errors could be very costly. Tape-preparation time was excessive; this resulted in long planning cycles and in idle time of expensive equipment.

Initial analysis showed that a three-fold program was needed:

1. A technique whereby a planner could effectively direct the operation of an NC machine tool in conventional planning terms with minimum translation of data from engineering drawings.
2. A device (probably a computer) to accurately perform the numerous, repetitive, routine calculations required to develop usable NC instructions. The device must also produce an operator's manuscript and punched paper tape for operating the machine tool.
3. A technique that would significantly reduce planning time and cost.

Initially, it appeared that the solution might exist in the development of a planning language and associated computer program. This would enable a planner to describe operation selection and sequences (the *creative* planning) while the computer did the routine calculation and decision-making (the *non-creative* tasks).

But this alone was not enough. Early PRONTO (PROgram for NUmerical TOol operation) programs were constantly undergoing revisions and modifications at considerable expense—to say nothing of the analysis, programming, and documentation problems.

Here was an opportunity for decision structure tables. They would provide a decision-making analysis technique, plus documentation, communication, and programming. Also, decision structure tables could be modified easily.

Thus, PRONTO was developed using decision structure tables. PRONTO is a generalized computer assist planning program for NC point-to-point machine tools. It also includes arcs, slopes and lines for milling operations. It is a "planner oriented" language that greatly simplifies the planning job. It allows the parts planner to "verbally machine" a part in his everyday planning language. Yet the language allows the most complex parts to be described easily with minimum drawing translation and calculation.

To produce a part on an NC machine tool requires two kinds of data:

1. Part-oriented or process data that relate to the part being machined.
2. Machine-tool-and-control-oriented data related to the specific machine tool and control being used.

Function of PRONTO Main Processor

Part-oriented data in the main processor include process description (drill, ream, tap, bore, etc.), dimensions, reference points, hole patterns, and tools. Basically, it is a geometric and machining description of the operations to be performed. It is these descriptions (locations, operations, and auxiliary functions) that will be converted into the information values and codes that will control the machining of the part.

Note that the part-oriented data are generally universal. The main processor, because it is process-oriented, can handle most parts as is. However, some modifications may be desired for special applications.

Function of PRONTO Post Processor

Unfortunately, machine tool and control data are *not* universal from one machine to another. Although recently significant progress has been made in the development of NC standards, most machine tools now installed were not built to these standards. This means that different machines have difficult codes, tape formats, and unique machine capabilities.

Therefore, the PRONTO "Generalized" Post Processor was developed. It was not written for a specific machine tool or control—but was written to include functional logic packages applicable to most point-to-point NC systems. It can be adapted to specific machine tools by selecting the proper routines (logic) and by completing specific information (data) applicable to the machine tool and control. This normally is about 75-80% of the logic and data. The remaining 20-25% required to complete a specific post processor must be developed for the specific application.

Typical post processor data include number of axes (X, Y, and/or Z), part reference or location on machine tool, tool or turret selection, modes of operation, auxiliaries, available feeds and speeds, etc., all specific machine tool and control data.

The planner specifies the location of each operation by specifying the X and Y coordinates of the center of the operation. That is the position of the tool in relation to a basic part reference point (0,0). For insance, the location of a drilled hole is defined by the X, Y coordinates of the center of the hole. In Fig. 2-1, X = one inch.

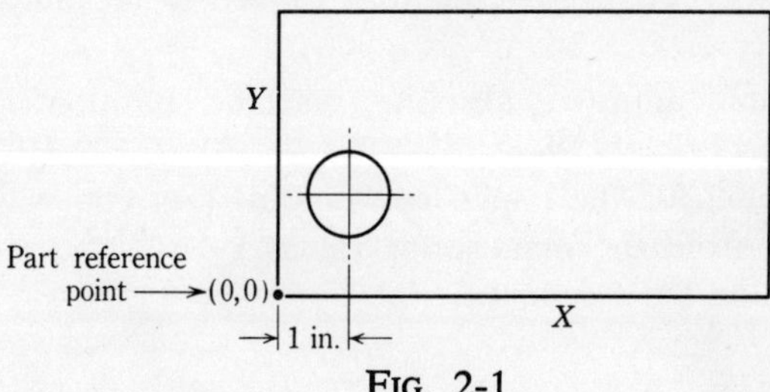

FIG. 2-1

The planner must also specify once for the part the relation of the part reference to the machine tool reference. This is the location of the part on the machine tool table. Now all dimensions can be related to the machine reference.

Thus, the X hole location is now six inches from the table reference. The next step is referencing to the spindle or absolute position, usually a function of the control.

The computer can calculate these current absolute values for all dimensions, in this case C-X.

These values must then be formatted for the particular machine tool and control being used. Two major control variations are "Memory" and "System." Memory can be YES or NO depending on whether the control retains a value until it is replaced (YES), or if the value must be repeated each cycle (NO). System is ABS, meaning the dimension is

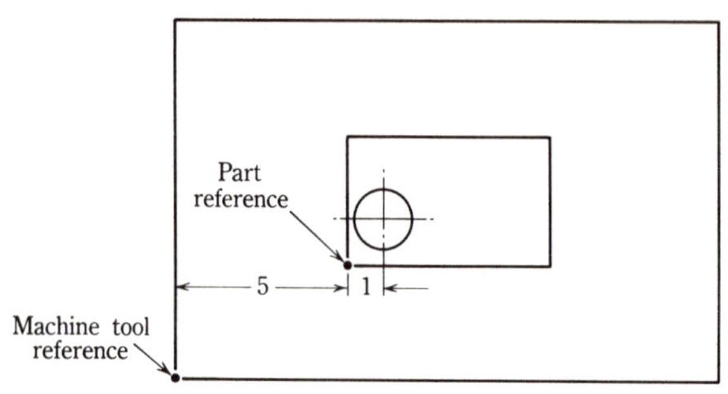

FIG. 2-2

the absolute position of the machine tool, or DEP, meaning departure if the next position is in relation to the last.

Two forms of output are obtained from the PRONTO postprocessor: Paper tape for the machine tool and a printed output for planners and operators. Here are additional definitions related to the output:

- M-X The present adjusted absolute machine position in the X dimension.
- PT-X The X value for the paper-tape output. (As you will see, this may vary from the manuscript values.)
- O-X The X value for the manuscript.

Table X-OUT-1 of the program now prepares the X output values for the paper-tape (PT-X) and the manuscript (O-X). Table X-OUT-DEP develops the X manuscript value for a departure system; the output value may be the absolute position or the departure value—depending on what information it is desired to give the operator. X-OUT-END is the end of the example.

TABLE X-OUT-1.

C-X	MEMORY	SYSTEM EQ	PT-X	O-X	GO TO
NEQ M-X	—	"ABS"	C-X	C-X	X-OUT-END
"	—	"DEP"	(C-X)-(M-X)	—	X-OUT-DEP
EQM-X	EQ "YES"	"ABS"	SPACES	SPACES	X-OUT-END
"	EQ "NO"	"	C-X	C-X	"
"	—	"DEP"	O	—	X-OUT-DEP

The next example will involve tape output format. Tape output formatting has been a continuing major problem. Variations are practically unlimited as these examples show:

Prefix Variations: Prefixes are one-letter designations a machine tool control uses to identify an item of data. For instance, X 12345 indicates the data 12345 are an X value. Some controls do not require prefixes; thus, the order of the data becomes significant. In this case, the items are usually separated by tabs (-). Successive tabs indicate no data for an item. The prefix variations of controls are numerous as shown by possible representations of the same value.

X 12345
-12345-
A 12345

Sequence Variations: The sequence in which the data are punched into the tape is another variable. The tape is read in blocks or groups of items. Each block controls a machine tool action. It may be positioning in the X and Y dimensions or tool positioning in the Z or depth dimension. The sequence of data varies within blocks for different controls.

One control may accept all the information in one block, another may require several blocks. Each of the following represents a valid output format—some requiring multiple blocks.

XY	(one block)
XYZ	
XY	(two blocks)
Z	
XY	(four blocks)
Z	
Z	
Z	

Digit Variations: Still another variation is the number of digits and right or left justification of the digits required for any item. Some variations of the value X = 10.500 are shown (b = blank).

010500
0105bb

b10500
b1050b

The table at the end of this example is the master output table where the manufacturing engineer originally establishes the tape format, order, etc., for any PRONTO postprocessor.

At this point, all the calculations, logic, etc., have been performed. It is now time to produce the tape. This table will determine the correct tape format for the machine tool, based on logic that the manufacturing engineer has supplied.

The following terms are used in the table:

LINE Some machine tool controls require several blocks of data to complete a motion or operation—each block represents a line of manuscript output—hence "LINE."

LINE 1 might contain the X and Y values, LINE 2 might contain the Z dimension.

ORDER This is order within a "LINE"—the first item, second item, and so on. In LINE 1 above, X was order 1, Y order 2, etc.

ITEM This is the specific variable whose current value will be output for the corresponding LINE and ORDER. Here the manufacturing engineer determines the tape format. In this example, ITEMS being output are SEQ-NO, a consecutive sequence number; X, Y, and Z position values; TOOL, the current tool identification number; AUX-1 and AUX-2, auxiliary functions such as coolants; and MODE, the tool motion sequence, a function of the operation. The X dimension is the ITEM for LINE 1, ORDER 1; the Y dimension is the ITEM for LINE 1, ORDER 2, etc.

TAPE FORMAT This is the output format that is desired for the ITEM. The first digit represents the number of positions to the left of the decimal, the second digit the number of places to the right of the decimal. For example, X = 12.345 would be of the 23 format, X = 12.34 would be the 22 format, and so on.

LEAD-SUPP TRAIL-SUPP Indicates by "YES" or "NO" whether suppression of leading or trailing zeroes is desired. Leading or trailing zeros are not required by some controls, therefore tape punching and reading time can be reduced by their elimination.

This table creates the output tape—the ITEMS are selected for proper LINE and ORDER. The TAPE-FORMAT and SUPPRESSION are determined next—the ITEM is then ready for tape output.

TABLE OUTPUT-1

LINE	ORDER	ITEM	TAPE-FORMAT	LEAD-SUPP	TRAIL-SUPP
1	1	SEQ-NO	40	"NO"	"NO"
1	2	X	23	"YES"	"NO"
1	3	Y	23	"YES"	"NO"
1	4	TOOL	40	"NO"	"NO"
1	5	AUX-1	30	"NO"	"NO"
2	1	SEQ-NO	40	"NO"	"NO"
2	2	Z	23	"YES"	"NO"
2	3	MODE	40	"NO"	"NO"
2	4	AUX-2	30	"NO"	"NO"

In our opinion, the use of decision structure tables in PRONTO helped:

1. Provide for an excellent analysis of the problem and an approach to the generalized solution,
2. Introduce the concept of "skeleton tables" whereby logic and relationship are furnished, but specific values must be supplied by the user. Skeleton tables include the condition and action headings as developed in the rules for writing structure tables. The values for the parameters must be furnished. For instance, different machines have unique feeds and speeds; thus the speed skeleton table may look as follows:

SPEED	SPEED	SPEED-CODE	SPEED-RANGE
GR	NGR		
GR	NGR		
GR	NGR		
GR	NGR		

Here the given speed, SPEED, is converted to the specific SPEED-CODE and SPEED-RANGE for this machine tool. Other machine tools may have different break points for the speed, or different codes and ranges.

3, Provide source documentation for manufacturing engineers, planners, and programmers.

4. Reduce programming time.

5. Minimize communications problems between users, specialists, and programmers.

6. Allow debugging at the source level (in English).

Thus the objectives of PRONTO were fulfilled largely through the use of structure tables.

WIDER ACCEPTANCE AND USE

Why isn't there wider acceptance and use of decision structure tables? There are three primary reasons.

The first is that structure tables are relatively new compared to flow charts. Most programmers began their training with flow charts and are understandably reluctant to make an abrupt change.

Second, there has been limited information available on how to do structuring. Earlier articles described what structure tables are without going into real "nut and bolt" explanations of how they are actually put together. Frankly, verbalizing the structure technique is difficult, as this article has demonstrated. Then too, relatively few persons have had actual experience with live problems.

Third, few structure table compilers are available today. Therefore, much of the benefit is lost because tables must then be coded. Experimenting with one of the existing processors probably isn't such a bad idea, since much of the advantage of structure tables certainly stems from the fact that they can be used directly to compile operating computer programs.

However, many benefits are not dependent on the computer. So, therefore, don't overlook the power of structure tables as a manual systems analysis and design tool. Indeed, this is where the principal benefits are claimed. Then too, if there is a mental or organization block about going someplace else to "compile" structure tables, you can implement a simple structure table of your own. They are remarkably effective and don't cost much. Simplified compilers ranging from $10,-

000–$25,000 have been built for a variety of machines. In comparison regular full scale pseudo languages are complex and may take hundreds of thousands of dollars to implement.

The important point is, don't wait for the world to provide a slick structure table processor. If needed, you can make one to satisfy your own needs on your own computer.

In closing, the authors would like to acknowledge the contribution of their associates in Advanced Manufacturing Engineering Service, Production Control Service, Computer Department, and X-Ray Department to the work reported in this article. Mention should also be made of the many contributions other General Electric departments have made to the development and application of decision structure tables.

This article, by D. T. SCHMIDT and T. F. KAVANAGH, originally appeared under the title "Using Decision Structure Tables" in *Datamation.* Reprinted with permission from *Datamation*, March, 1964, published and copyrighted 1964 by F. D. Thompson Publications, Inc., 35 Mason Street, Greenwich, Conn. 06830.

3

Engineering Data Processing Using Decision Tables

How do we picture the automated engineering operation of tomorrow?

Can you picture how design and industrial engineers will be able to concentrate their efforts on the creative and logical processes and not have to be concerned with the physical paper work manipulation? Can you see the massive disk files recording every bill of material and every planning record that's ever been used? Can you visualize the rapid terminal-oriented information retrieval system needed to get up-to-the-minute information and facts? Can you picture as automated a drafting system using light pen input on graphic terminals to do the necessary recording of the shapes and curves with automatic print-out in hard copy form?

It's quite a picture, indeed. The process sounds believable—all the equipment is available today to do the job. Yet I predict that the system envisioned above is not an accurate portrayal of the automated engineering system of the future. Why not? Because there is a way to avoid the massive files, the elaborate information retrieval and the complex blueprint making.

THE DECISION LOGIC CONCEPT

Instead of taking the usual approach of simply automating the previous manual procedures, a new concept is needed. This concept is called decision logic regeneration.

What does it mean? Simply this—instead of storing previously developed answers and retrieving them on demand, it is often less expensive and more efficient to store the decision logic that was used in developing these answers and regenerate the answers each time they are needed. The present state of computer technology makes the economics of decision logic regeneration extremely attractive. From a systems viewpoint, the business operation can be performed at less cost with higher quality and greater response time than with the usual file retrieval approach.

Design engineering and manufacturing engineering are two of the major areas in which this decision logic regeneration can be profitably applied. Unfortunately, the nature of these applications requires extensive computer programs with relatively complicated logic; yet, there remains a need for easy maintenance and revision. The usual techniques of narrative description, flow charting, and assembly language programming become uneconomic for this type of application. It has been necessary, therefore, to develop additional techniques to permit the efficient installation of these regenerative design engineering and manufacturing engineering systems. One special technique which enables the data processing team to grasp and record the existing design or planning logic is used in practically every step of the process—from documenting the current engineering logic to preparing the necessary computer programming. This tool is decision tables. Decision tables are a natural outgrowth of previous work on structuring and recording related information.

(*Editor's Note:* Burton Grad's discussion of the basic elements of decision tables has been omitted at this point.)

Design Engineering

With this general background on decision tables, let's see how they can be profitably applied to design engineering.

A decision table skeleton for an engineering system is shown in Table 3-1. Customer specifications are shown in the condition area of the table and product characteristics appear in the action area below. Names of the specifications and names of the characteristics are shown in the stub, while various values for those names are displayed in the entry area.

A filled out engineering decision table is shown in Table 3-2. Various customer specification are listed on the left: service, application, etc. These are the conditions for this example. Below the third horizontal line and to the left of the vertical line are the resulting actions or product characteristics: type of armature, number of windings, etc. To the right of the vertical double line are the values for the customer specifications

TABLE 3-1. STRUCTURE OF DESIGN ENGINEERING DECISION TABLE

STUB	ENTRY
Customer Specification Names	Specification Values and Ranges
Product Characteristic Names	Characteristic Values and Ranges

and product characteristics. A decision rule is a unique combination of specifications and characteristics.

TABLE 3-2. BASIC DATA

Table No. 1	Rule 1	Rule 2	Rule 3	Rule 4
Service	DC	DC	AC	AC
Application	Temperature	Speed		
Rating Units	MV	MV	MV	MA
Number of Phases	1	1	1	1
Type of Armature	Moving Coil	Moving Coil	Electro-dynamic	Inductive
Number of Windings	1	2	2	1 + Number of Phases
Part No.	012526	012526A	012530	012535
Assembly Drawing No.	A26	A26A	B30	B30A
Next table	2	2	2	10

For example, rule 1 reads: IF the service is DC, and the application is temperature, and the rating units are MV and the number of phases is 1, THEN the armature is a moving coil type and the number of windings is 1 and the part number is 012526 and the assembly drawing number is A26 and the next table is 2.

In automated design engineering, decision tables are used to:

1. capture the engineering logic;
2. review this logic for consistency, accuracy, and technical correctness:
3. simplify the decision rules for most effective communication;

4. represent in programmable form the design logic so that programs can be written from them readily or some type of automatic conversion made;

5. retain the decision logic in a form which permits efficient revision as engineering design changes take place.

For the complete design of a product like an electrical instrument, transformer, pump, or motor, many decision tables (often thousands) may be required. Part of the logical process is to relate these tables to each other so that in performing the total product design, each table is entered correctly and intermediate results are saved and used in subsequent decision tables.

Manufacturing Engineering

Let's look at a separate engineering problem, that of manufacturing engineering or planning. In this area, the intent is to capture the logic used by the manufacturing engineer in determining exactly how to manufacture a particular part. There are three types of problems in manufacturing planning for which decision tables are directly applicable: routing, operations methods, and time standards.

Let's look at a problem for sheet metal routing to illustrate how such a problem would be solved. Fig. 3-1 shows the decision table. These sheet metal parts may or may not have holes or bends. The parts which have both holes and bends will usually have the holes perforated on a punch press and then have the bends formed. If threads are specified, holes are drilled and tapped after the forming operation; otherwise, the holes may become deformed during the forming operation. If the hole diameter is equal to or less than the thickness of the piece, the hole is drilled instead of being perforated. Note that the condition stub holds the parts characteristics and the action stub the machine assignments.

In a similar way, the operations method logic would be captured covering the areas of tooling, gross material selection, machining details, and operator instructions. Note that a set of decision tables is able to handle not just a single part, but an entire family of parts. The set can handle not just a single station, but a whole area of the factory.

System Piece parts

Page 1 of 1

DECISION TABLE

40 Sheet metal routing	1	2	3	4	5
Holes	Y	Y	Y	Y	Y
Threads specified	N	N	Y	N	N
Hole dia. > thickness of mat'l	Y	N	–	Y	N
Bends	N	N	N	Y	Y
Perforate holes (punch press)	X			X	
Form (punch press)				X	X
Drill holes (drill)		X	X		X
Tap holes (drill)			X		
Go to Table 41	X	X	X	X	X

FIG. 3-1. AUTOMATED MANUFACTURING PLANNING

Programming

Capturing the design and planning logic is only part of the job. It is certainly a big part of the job and requires careful coordination between the data processing man and the engineers and planners. But, just having these "English language" decision tables does not make them operational on a computer. To do this requires an effective programming effort. These are often very substantial jobs. For instance, many products will require 50,000 instructions or more for design engineering and a similar number for manufacturing planning. But the nature of the problem is such that this should not be an overly expensive task.

The techniques of program conversion have been well worked out in the past few years. There are four principal approaches which can be used: manual, semi-automatic conversion, interpretation, and automatic conversion.

Within each of these possible approaches there are certain techniques used which can make the conversion more convenient and make the operating program both more compact and efficient in running time.

The manual approach involves re-writing each decision table for more efficient and compact representation. For instance, the most frequently used rule may be solved first and the most likely to fail condition may be examined first. Other techniques would include grouping identical conditions and establishing patterns of associated actions. Programmed switches are often used to represent the results of complex tests. Various path controlled mechanisms are useful in the action area. In general, manual programming from decision tables is convenient and leads to reasonably efficient object programs. The programs themselves may be written in an assembly language or any particular higher level language.

Semi-automatic conversion uses the computer to assist in analyzing the table structure and in converting from the user source language to an acceptable computer input language. Normally, some attempt will be made to combine like conditions and handle common actions on a consolidated basis. Problems arise in the choice of source language and the complexity of the analysis performed to achieve solution efficiency and program compactness. The output from such a program will normally be in a higher level language so that manual changes and correctness can be made. The 1401 Decision Logic Translator is a program of this type which goes from a FORTRAN-like source language to a legitimate FORTRAN program.

The interpretive approach often has much to recommend it since it allows for direct storage of the decision table, usually in a coded or compact form, thereby insuring easier maintenance. The disadvantages include the necessity for having a special interpretive program in core memory at all times and the usual slower solution speed of an interpretive approach. A number of such programs have been written by various individual companies. Most of these programs have had to set some limit on the number of rows and columns in the table and have had to establish fairly restrictive rules on what constitutes a proper stub and a proper entry. The interpretive programs have been relatively small—often in the order of 300 to 500 instructions. If maintenance is a key concern, this may be the best approach.

Automatic conversion programs are those which will accept decision tables written in a user source language and completely convert them to fully acceptable computer input, usually at the machine language level. Most programs written to date of this type have gone through an intermediate language such as FORTRAN or a particular assembly language in order to reduce the total costs of writing the compiler. The difficulties with this approach have been two-fold: cost of the compiler is high and the efficiency and compactness of the resulting operating program is often questionable. This latter point is caused by the complexity of the logic needed to generate compact, efficient programs under all conditions.

The Future for Engineering Use of Decision Tables

It is our feeling that the growth of decision logic generation for both design engineering and manufacturing engineering will be very rapid. As a new product is designed or a new factory planned, the basic logic will be recorded in decision table form for growth and development with the product line. Engineering departments will become familiar and comfortable with these techniques and train their people in their use. The impact, in terms of lower product cost, greater turn-around speed, design standardization, and flexibility of response will radically change the present relationship of engineering costs to engineering values.

This article, by BURTON GRAD, appeared in *Data Processing, Vol. VIII* (the Proceedings of the 1965 International Data Processing Conference). It is reprinted here with permission of the Data Processing Management Association, Park Ridge, Ill.

4

The Value of Decision Tables in Manufacturing

To date, the difficulties of communicating with electronic computers have received much attention. The various pseudo-languages represent great advances in this area, but a language is a great deal more than the basic tool of communication. A good language, a good symbology, is an essential element in man's thought processes. In a sense it defines his capacity for conceptualization and abstract thought. Today we face great language restrictions in trying to analyze and think about the complex decision-making systems required to operate a business or control an industrial process. Our traditional techniques seem inadequate. Flow charts quickly become a puzzle of lines, balloons, and boxes whose secret lies hidden in the mind of the creator. Frequently, programmers complain they would rather reprogram the job than take over someone else's flow charts.

In addition to flow charts, we often see matrix-type displays which appear under a variety of names, such as collation charts, tabulated drawings, standard time data sheets, and so on. Often large and unwieldy, they usually represent listings of past decisions or answers rather than the logic used in making them. But none of these methods for thinking about and communicating complex decision logic has been really effective. Most business and professional people must still communicate with the computer through an elaborate hierarchy of flow charters and programmers. We think that structure tables will shorten this chain, allowing more direct communication between the ultimate user and the computer. The structure table approach combines some of the characteristics of these earlier methods and introduces a few new features of its own. After using decision structure tables in numerous applications, we feel they are not only good communicating tools but also valuable thinking tools.

(*Editor's note:* The Authors' discussion of decision table fundamentals has been omitted at this point.)

Applications

We now have a rather substantial amount of experience in applying structure tables to a wide variety of operating decision-making problems, and we would like to focus on some of these applications, particularly within manufacturing.

To appreciate these applications, it would be well for us at the outset to share a common understanding of just what manufacturing is, as opposed to marketing, finance, engineering, employee relations, etc. In general, manufacturing converts marketing and engineering specificatios into finished products ready for customer use; it buys tools and materials, it runs factory machines, it assembles parts, it tests and inspects products, it packs and ships them to customers. More than just doing the actual work, manufacturing also concerns itself with design engineers who are interested in the functional soundness and appearance of products, so also there are engineers in manufacturing who concern themselves with:

- What is the best machine or process?
- How fast should it run; what tools should be used?
- How can we be sure that the parts are good?
- When should we make the parts, how many?

All of these questions, and thousands more like them, are the *everyday province of the engineer in manufacturing.* Getting better answers to these questions means more efficient shop operations, lower manufacturing costs, and improved values for customers. This is the work of manufacturing.

MANUFACTURING VIS-À-VIS COMPUTERS

There are three factors in manufacturing's relationship with computers that you should know in order to appreciate their application. First, computers are still relatively new to the manufacturing function.

Where employed, the computer has been used in large measure to perform routine clerical operations such as filing, sorting, printing, and the like. Rarely, for example, does the computer enter into manufacturing decision making. The reasons are manifold. The tremendous volume of information and the many complex, detailed interrelationships have made it extremely difficult and costly for manufacturing people to formalize their logic. Manufacturing still relies heavily on "experience" and "art" as opposed to explicit analytic procedures and quantified design techniques.

Second, computer hardware development is only now beginning to provide the size, capability, and cost which manufacturing needs to install computers at attractive cost reductions; that is, numerical methods using computers are now beginning to come up with better, cheaper, faster answers than the "artisan."

Third, today's manufacturing man knows very little of electronic computers and even less about programming them. If computers are to make real inroads, we must find direct, practical ways to show the manufacturing man what the computer can do for him, and we must also develop efficient methods whereby he can learn to use them himself. We cannot train enough programmers to code the problems that exist in the manufacturing function. Even if we had the money, our human resource would fail.

From this introduction you can gather the fundamental appeal which decision structure tables have for manufacturing. They provide a simple, uniform method for describing complex decision systems. The precise, unambiguous format facilitates technical communication within the organization and provides a formal documentation procedure. This is becoming increasingly important in these days of multi-functional integrated computer systems. To be specific:

1. Structure tables provide a disciplined decision analysis which forces consideration of logical alternatives. The precise format highlights illogical statements and emphasizes causal relationships. It constantly directs attention to why things are different. Intelligent standardization can be fostered.
2. Structure tables are easy to understand and formulate. The tabular format is a reasonably familiar language form. It is not a tremendous departure from the tables the manufacturing man has already used in methods planning, time standards, lot size determination, sampling, and so on. Further, the format is so simple and straightfor-

ward that manufacturing specialists can write structure tables for their own decision-making problems with very little training. This cuts computer application costs and cycle times.

3. Debugging is simplified. Structure table errors are reported at source-language level, thus permitting the specialist to debug without a knowledge of detailed computer coding.
4. Lastly, structure tables are easy to maintain. Instead of changing all the precalculated answers in all files, it is often necessary to change only a single value in a single table. In this way the computer is always in position to calculate up-to-date answers.

MANUFACTURING APPLICATIONS

Thus the manufacturing applications problem is really not one of verb versus verb, or microseconds matching microseconds. At this time, we are concerned primarily with making manufacturing aware of the many practical things computers can do. Our problems consist of demonstrating technical feasibility, proving economic value, defining problems, organizing and maintaining large amounts of data, training people, and so on. Our experience indicates that decision structure tables can really help us in this endeavor. These applications on rotors, gears, and inventory control provide some reasons for our belief.

Cast Rotors

General Electric has an understandable interest in electric motors. In one of the earlier manufacturing structure table applications projects, a study was made of the centrifugal casting process used to make rotors for a line of alternating current motors. As you recall, electric motors consist of two basic parts: a stationary frame or stator, and a rotating element or rotor. The rotor was made from slotted steel laminations, and copper wire was wound in the slots. The wires were connected at each end to form a complete electrical circuit. When the rotor was placed under the energized electromagnetic poles on the stator, torques were set up which made the rotor spin around. The basic theory of electric motors hasn't changed very much; however, if you take apart your washing machine you will very likely find that the copper has been replaced by aluminum. Moreover, the aluminum is not wound in the slots, but rather the rotor has been molded together as one solid piece. In addition, odd-shaped protrusions may be sticking out from the end. These fins serve as fan blades for cooling the motor. Many of these cast rotors are made using a centrifugal casting process; that is, the mold is rotated at high speed so that the liquid aluminum metal will be forced eventually into all rotor slots, fan blades, and other crevices in the mold. In addition, spinning also helps to prevent the formation of bubbles and voids in the aluminum itself.

In the particular line of cast rotors that were selected for study, over 100 varieties were currently active and, of course, new varieties might appear at any time. The differences in rotors were basically caused by different design techniques and configurations used to cool various horsepower motors. To factory operators, this involved different assembly procedures in putting together molds and rotor laminations. In addition, depending on a number of other variables, there were also differences in the detailed casting procedure.

The first step in the structure table development project was to extract these methods and procedures as well as their supporting logic from widely scattered sources in the current manual system. A considerable portion of the required information existed only in the minds of people doing the job. The results were summarized in approximately 60 decision structure tables. These tables covered not only the 100 varieties then active, but also provided planning logic for some 44,000 rotor configurations then possible.

In addition to describing the factory operating procedure, these structure tables also developed the time standards, the "allowed" or normally expected operation times. The resulting computer program printed out labor vouchers and also detailed factory operator instructions which told shop people how to build each rotor. The entire project was completed in six weeks by a man who was then unfamiliar with structure tables, computers, or the rotor casting process.

Subsequent to the completion of this work, it was decided essentially to re-do the project with a new man using flow charts and what were then more conventional programming techniques. This would make about as controlled an experiment as is possible in an industrial environment. One cannot really generalize from a single observation, but perhaps you might be interested in a few comparative statistics. The second project took 14 weeks in contrast to six. The second computer program produced similar output, but required a 50% larger object program than was developed using structure tables. However, the structure table program ran one-third slower. The size and speed differences were largely due to different approaches to computer implementation. It was clearly demonstrated that both approaches had their merits, but 14 weeks versus 6 weeks really seemed to be a good omen.

Gears

Many General Electric departments share a common interest in the production of gears. This component appears in hundreds of the company's products. However, many are unaware that the common gear is an uncommonly complex thing to produce. Indeed many engineers —and some entire companies—devote all of their interests to the proper maufacture of just this one type of component.

While some simple gears can be molded out of plastic, the more substantial variety in which we are interested is typically made from flat round metal discs called blanks. Typically, these blanks are forged individually or cut from lengths of bar stock. In general terms, an average gear might be manufactured as follows: First the blank is rough machined, front and back, to provide a gripping surface; this also eliminates scale and some excess material. Then a center hole might be drilled, bored, and reamed to provide a concentric, perpendicular locating surface for subsequent machining operations. Perhaps a key way or a spline will be formed with a broach inside this center hole. Once these operations are completed, the gear may then be finish-machined, front and back, giving the web and flange its final shape. It is only at this point that the gear is ready for hobbing, one of the conventional processes used to cut teeth. Following this, the teeth and other parts of the gear are ground to provide a smooth surface and close dimensional tolerances. In between these operations frequent trips to the annealing furnace are required to relieve internal stresses which machining has set up within the metal itself. A survey of machined parts shows that the average part goes through five operations; gears average around 30. Gear manufacturing can be a complex job, to say the least.

In this applications project, the task was to write decision structure tables to completely describe the operation planning for all factory operations in a large family of complex gears. This is the task of determining which machines will perform what metal working operations, in what sequence, with what speeds and feeds, with what dimensional tolerances, what tools, and within what time. The results of these decisions were to be furnished to factory operators in the form of printed instructions which contained enough detail to permit them actually to make the gears in question.

At the time the tables were written, this family of gears was already several hundred strong; however, the objective was to automate the planning for expected additions as well as to simplify file maintenance and clerical operations on the gear planning already in existence. The project to write and debug the 3,000-odd structure tables which resulted took approximately two and one-half man years. The resulting computer program contained over 60,000 instructions. The structure tables were written by manufactuirng planning technicians who knew little about the world of computers. As a result of their endeavors, the cycle for planning a new addition to this gear family was reduced from 4 weeks to 20 minutes of GE 225 computer time. The use of decision structure tables greatly facilitated documentation of the logic and uncovered many opportunities for standardization. Functioning in much the same fashion as engineering blueprints, the decision structure tables have now become the official manufacturing engineering documentation of the work. The program is working now and is expected to break even in the first six months of operation.

Inventory Control

In a completely different type of project, decision structure tables were used to develop TRIM, a computer simulation model to test rules for inventory management. Here structure tables were used to describe statistical forecasting and automatic ordering rules, as well as a simulation language. TRIM was developed to provide a controlled environment laboratory where systems designers could experiment freely with a variety of inventory control decision rules without disturbing the real world. TRIM, like most computer simulation programs, offers three major advantages. First, TRIM operates much faster than real time. It can simulate 50 time periods of inventory systems activity in three to five minutes. Second, because it is a computer model, it is possible to explore extreme situations without risk of destroying the model or, perhaps more important, the actual inventory system itself. Third, computer simulation provides a controlled experimental environment where cause and effect relationships can be established with a much higher degree of certainty than can ever be done in the real world.

TRIM essentially makes a GE 225 computer behave like a complete single-stage inventory system. It processes customer demands, estimates future requirements, places and receives replenishment orders, purges overage inventory, cancels over-extended back orders, etc. TRIM also reports how well the inventory decision rules succeed in balancing customer service, ordering costs, and inventory carrying charges in accordance with specific weights the user attaches to these measures of performance.

TRIM is controlled by a so-called "master clock." Each significant activity, forecasting, ordering, etc., is assigned a separate alarm clock. The function of this alarm clock is to let TRIM know when this particular activity or transaction will occur *next*. The alarm clocks are carefully sequenced so that if there should be a tie with two or more alarm clocks going off at the same time, TRIM will handle the activities in proper logical as well as chronological order. The master clock constantly records "current time" and coordinates all subordinate alarm clocks. Within each activity, there is usually a number of options available to the user. For example, in forecasting, TRIM already contains moving averages, single, double, and triple smoothing. Also included are a user controlled "panic" policy as well as an independent control for trend extrapolation. Lead times for replenishment orders may be fixed by input or determined stochastically from a distribution function. Or if the user wishes, lead times can be established by using a small factory flow shop already built into the model.

TRIM was developed completely in decision structure tables. Tables were used to describe the master clock, to generate random numbers, to search and housekeep lists, to update records, and last, but not least, to make inventory decisions. The model is now written in GECOM-TABSOL,* which is currently in field test in customer installations. It contains over 100 decision structure tables which produced approximately 7,100 words of programming. Start to finish, TRIM takes 45 minutes to compile. It runs on a minimum configuration GE 225 card input with on-line printer or punch. The original program was completed by a two-man team in about three months. The program is operational and has been successfully used by a number of General Electric product departments to analyze existing or proposed inventory systems. Results indicate 20 to 30% reductions in average inventory, with the same or improved service, not to mention lower clerical costs and reduced errors. The TRIM program is now available as an official 225 software package.

* Discussed in the conclusion of the paper.

Computer Implementation

Thus far, the discussion has emphasized the *application* of decision structure tables. In order to discuss more fully the use of structure tables, however, it is also necessary to consider some questions pertaining to *computer systems design* because much of the benefit of the tabular approach would be lost without a good sound computer implementation. In particular, three of the problems which are of major concern to the systems designer are:

1. Need for flexibility.
2. Dynamic memory requirements.
3. Extensive amounts of list processing.

We would like to describe some of the ways in which these problems have been met in our structure table applications. The techniques to be described are neither revolutionary nor are they peculiar to tabular languages, but they do represent significant advantages over some earlier efforts in this area and have greatly enhanced the usefulness of decision structure tables.

For the sake of definiteness, our discussion will be based upon the use of structure tables in simulation programs such as TRIM. However, the problems and their proposed solutions are quite generally applicable, and the reader should have no difficulty in translating the word simulation into one which more adequately reflects his particular area of interest.

Relocatable Tables

Because a computer model is used to evaluate many different alternative decision systems, the program must be flexible so that changes in the logic can be incorporated with a minimum of reprogramming and recompilation. Our approach to this problem has been to develop modular simulation programs built upon completely relocatable structure tables.

Structure tables compiled in 225 TABSOL are relocatable in the same sense that FORTRAN II subroutines are relocatable. Absolute memory assignments are not made at compilation time. Instead, memory is assigned only at execution time by a specially designed "load program"

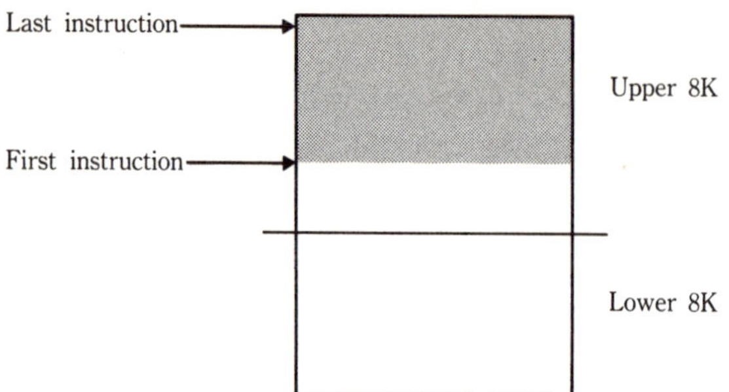

FIG. 4-1. RELOCATABLE TABLE

in such a way that the last instruction of the program will be located at the very top of memory. That is illustrated in Fig. 4-1. The reasoning behind this approach will become clear when the 16K memory layout is discussed in the next section. *Any memory not required by tables becomes available for data storage.*

Reduced Compilation Time

One of the principal benefits of relocatable structure tables is that all tables need not be recompiled to incorporate new problem logic or to make corrections. Instead only those tables which have been modified need be recompiled. For a practical illustration of what this means, we might consider the TRIM program. TRIM consists of about 100 structure tables. Without relocatable tables, making a modification to any *one* of these tables required a complete recompilation of all 100 tables. The total compilation time was about *40 minutes,* or an average of 0.4 minute per table. With relocatable structure tables, however, only the table to be modified need be recompiled. Allowing setup, the average compilation time per table is not more than *2 minutes.* Thus, compilation time has been reduced by a factor of 20 to 1! It makes no difference if the recompiled tables are longer or shorter than their original versions. At execution time, the loader will adjust the origin of the tables downward or upward by just the right amount.

A Library of Tables

This feature makes it possible to build a library of tables which can be put together at execution time to provide a custom-tailored program. For example, suppose that a sample program consists of three sequential decisions and that each decision part requires one structure table. Furthermore, assume that there are five alternate tables (versions) for each decision and that any table for the first decision can be used with any table for the second decision which, in turn, can be used with any table for the third decision. Thus there are 125 ($5 \times 5 \times 5$) different computer programs which can be put together from these 15 tables. *Without* relocatable tables, it would require 125 compilations of three tables each or the equivalent of 375 single table compilations to achieve these 125 unique programs. *With* relocatable tables, each table need be compiled only once, giving a total of 15 single table compilations. Thus, the relocatable feature reduces compilation requirements by a factor of 375/15 or 25/1 for this simple illustrative example.

Efficient Use of Memory

Any table that will not be required for a given run can be omitted from the program deck *without substituting a dummy table as its replacement*. In this respect, the new 225 TABSOL differs from the relocatable feature of FORTRAN II. Let's take a simple example. Suppose Table 1 can GO TO either TABLE 2 or TABLE 3 depending on the value of some calculated parameter. Further suppose that the input for the next run is such that TABLE 1 should never branch to TABLE 2. If TABLE 2 is a large table, it may be desirable to remove it from the program in order to make additional memory available for data storage. In 225 TABSOL, we simply remove TABLE 2 from the program deck and let the loader do the rest.

FLEXIBLE 16K MEMORY LAYOUT

A 225 TABSOL program consists of two parts:

1. The compiled table logic.
2. The executive and library routines.

As mentioned above, table logic is pushed upward to the very top of memory by the special load program discussed in the previous section. The TABSOL executive and library routines are placed at the very bottom of lower memory.

All memory locations between the end of the TABSOL subroutines in lower memory and the origin of the table logic in upper memory are available for *common* data storage (Fig. 4-2). Thus there is no preassigned limit on the number of parameters in a 225 TABSOL program.

The limit depends entirely on how much storage is required by the program. Finally, we would point out that all data are automatically stored in this common area; there is never a need to use arguments in order to communicate between tables.

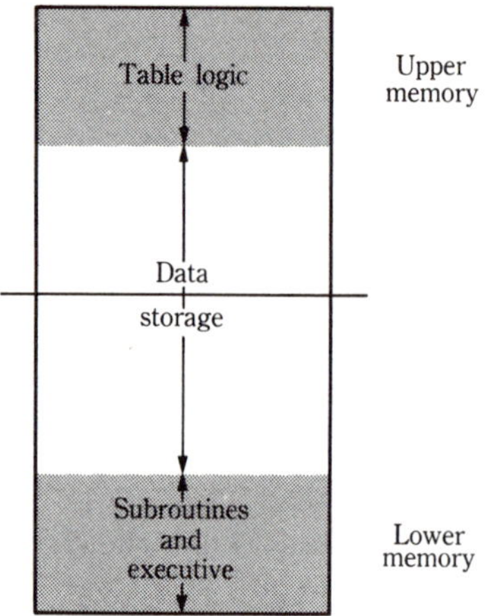

FIG. 4-2. COMMON DATA STORAGE

Divisions of the Data Area

In 225 TABSOL, there are two types of data:

1. Non-record variables.
2. Record variables, sometimes called record fields.

Physically, non-record variables will be located contiguous with the end of the subroutines in lower memory, as shown in Fig. 4-3. The record area of memory extends from the end of the non-record area to the origin of tables. The end of the non-record area is determined by the number of non-record parameters used.

The origin of the table logic is established by the loader, as indicated above. Furthermore, the loader will store this origin as a specified parameter in the non-record area. The location of the table logic origin is an *essential* piece of information since the executive routine must have some way of determining where to transfer control.

The starting location of the table logic is a desirable feature from an entirely different point of view. In programming applications such as

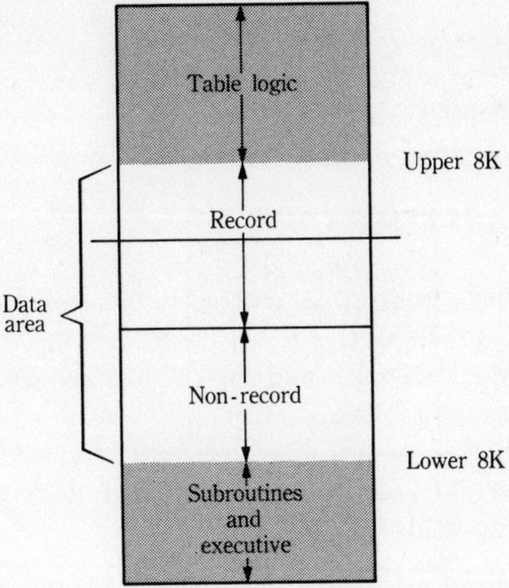

FIG. 4-3. NON-RECORD VARIABLES

simulation, the amount of working data storage required cannot be predicted in advance because it depends on what happens during the running of the program. Because the origin of the tables has been recorded in the data area, the programmer can now process this variable just like any other parameter in the data field. In particular, when more storage is required by the events of the program, he can easily calculate whether such storage is available.

RECORD PROCESSING

Computer programs usually require a considerable amount of list maintenance. There are two problems here:

1. Efficiently presetting the dimension of each list is difficult.
2. The extensive amount of subscripting required to manipulate these lists tends to confuse and cloud the essential logic of the program.

As a partial answer to both these difficulties, a record processing technique was developed.

Properties

A record has two basic properties:

1. A name
2. An ordered set of fields.

The *name* identifies the record. For example, we may refer to an *ITEM* RECORD, or an *ORDER* RECORD, or a *MAN* RECORD, etc., where ITEM, ORDER, and MAN are names which identify three different types of records.

In general, each type of record will differ in its *ordered set of fields* (the second property). For example, an *ITEM* RECORD might consist of the following fields.

FIELD NAME	FIELD NUMBER
NEXT	1
ONHAND	2
ONORDR	3
FORCST	4
DEMAND	5

An ORDER RECORD, on the other hand, might consist of an entirely different set of fields, as, for example:

FIELD NAME	FIELD NUMBER
NUMBER	1
BUYER	2
QUANT	3
DATE	4
DESTIN	5

Number of Records of a Type

In general, there will be more than one record of a given type. If this were not the case, there would be no need to define the type as a record in the first place. For example, suppose that we are writing a program to control inventories at a warehouse which stocks seven different items. In this case we would need seven ITEM RECORDS, each of which contains the same fields (NEXT, ONHAND, ONORDR, FORCST, DEMAND, in this example).

Size of Record Type

The size of the record type will be equal to the number of fields which have been defined for the record. Further, each record of a given type will have the same size. Thus, in our example, the size of an ITEM RECORD is five. The only exception occurs when the record contains a list. For example, assume that we wish to store DEMAND for each item during each of the last 12 months. In this case, the size of an ITEM RECORD would *not* be five. Instead, the size would be sixteen (NEXT + ONHAND + ONORDR + FORCST + DEMAND (1) + DEMAND (2) + ... + (12)).

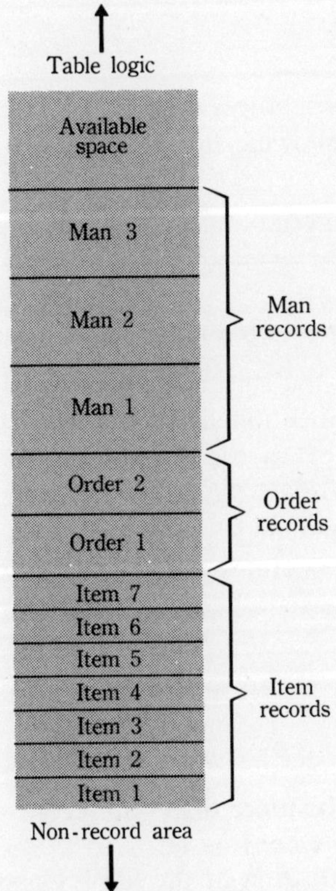

FIG. 4-4. RECORD LAYOUT

Layout of Record Area

The record area will be organized so that all records of a given type will be contiguous and sequenced by their record number. For example, if there were seven ITEM RECORDS, then the record for item 1 would record for item 3, etc.

Figure 4-4 illustrates what memory might look like if we had 7 ITEM RECORDS, 2 ORDER RECORDS, and 3 MAN RECORDS.

THE IMPORTANT POINT TO NOTE IS THAT THE NUMBER OF RECORDS OF ANY TYPE AND TO A CERTAIN EXTENT THE SIZE OF A RECORD OF ANY TYPE IS COMPLETELY CONTROLLED BY INPUT. In other words, to increase the number of ITEM records from 7 to 10, for example, does not require recompilation, nor does it require setting aside any storage in anticipation of these three additional items.

The GET Command

The vehicle for record processing in 225 TABSOL is the GET Command, which has the following format:

RECORD	
GET	"NAME"

The word RECORD in the stub row appears only for the sake of readability and serves no other function. The word GET is, of course, the operation code. The key to the functioning of this command is the "NAME" which appears in the related parameter field, and this is going to take a little explaining.

The GET Command operates in two modes depending on what kind of "NAME" is used.

- Mode 1 operation occurs when the "NAME" is a parameter in the non-record area.
- Mode 2 operation occurs when the "NAME" is a field of some record type.

Let's illustrate the operation of the GET Command in *Mode 1*. Assume the working variable ITEM has been defined as a non-record parameter and that the record for item 5 is to be processed. The procedure is as follows:

ITEM		RECORD	
PUT	000005	GET	ITEM

Subsequent to this point, if we refer to ONHAND, for example, it will be the ONHAND for the 5th item. Thus, suppose we wish to subtract 10 from the onhand for item 5. Having already "gotten" the record for item 5, we very simply:

ONHAND	
SUB	000010

Now let's take a little closer look at what the GET Command did. Not only did it get a particular type of record, namely an ITEM RECORD, but it also got a particular ITEM RECORD, namely that for item 5. The point is that the parameter ITEM serves two functions:

1. It identifies a type of record.
2. It identifies a particular record within that type.

This is not the case, however, when the "NAME" used in the GET Command is itself a record field. Thus, in *Mode 2* the "NAME" will not identify a type of record. It will identify only a particular record of the type of which "NAME" is a field.

This really isn't as confusing as it sounds. As an example, suppose that we wish to compute the total value of inventory on hand. Let this be called TOTAL, a non-record parameter. Now TOTAL is simply the sums of the ONHAND's for each item. Therefore, to compute TOTAL, the program must in effect loop through all ITEM Records, noting the ONHAND, and accumulating into TOTAL.

To facilitate this "looping," we can set aside a field in each ITEM RECORD called NEXT. Further, we can set the value of the NEXT field as follows:

ITEM	VALUE OF NEXT
1	2
2	3
3	4
4	5
5	6
6	7
7	0

The program is shown in Fig. 4-5.

TABLE 1

ITEM	RECORD	TOTAL	GO
PUT 000001	GET ITEM	ADD ON HAND	TABLE 000002

TABLE 2

NEXT	RECORD	TOTAL	GO TO
GR 000000	GET NEXT	ADD ON HAND	TABLE 000002 TABLE END

FIG. 4-5. PROGRAM ILLUSTRATING "GET" COMMAND

In other words, the

RECORD	
GET	NEXT

Command simply gets the next ITEM Record. Thus, the value of NEXT in the first ITEM Record is 2. Accordingly, having "gotten" the first ITEM Record in TABLE 1, the command

RECORD	
GET	NEXT

will get the second ITEM Record, and so forth.

Variations

The reader who finds Mode 2 operation confusing might wish to take advantage of the fact that the word RECORD in the stub is quite arbitrary. Thus instead of saying

RECORD	
GET	NEXT

it is *equivalent* to say

ITEM	
GET	NEXT

In other words, the stub is ignored when the GET Command is used. Accordingly, the stub can contain anything which makes the Command more comprehensible.

ASSOCIATIVE LIST TECHNIQUES

Because of the dynamic storage requirements of simulation programs, associative techniques for manipulating lists are frequently useful. We would like to illustrate how the language would handle such a situation.

As a test case let's consider the problem of filing an order with a certain priority, PRIOR, into a queue, QUEUE. Those orders which are already in queue we assume to have been previously sequenced according to their respective priorities. The objective, therefore, is to insert the new order into the queue so that the correct sequence is maintained. We assume that the lower the priority number, the more urgent the job.

Let us assume that the layout of an ORDER RECORD is as follows:

First Word	IDENT	(Identification no.)
Second Word	PRIOR	(Priority)
Third Word	PREDEC	(Record number of Predecessor order)

Fourth Word	SUCCEC	(Record number of Successor order)

We shall adopt the convention that the first order in queue shall have a predecessor of zero and that the last order in queue shall have a successor of zero.

There will be a record type called QUEUE RECORD in which the following information, among other things, will be stored.

FIELD	FIELD NUMBER
FIRSTO	1
LASTO	2

FIRSTO and LASTO note the record numbers of the first and last orders in this queue, respectively. The initial state of the ORDER RECORDS will be as shown in Figure 4-5. The successor and prede-

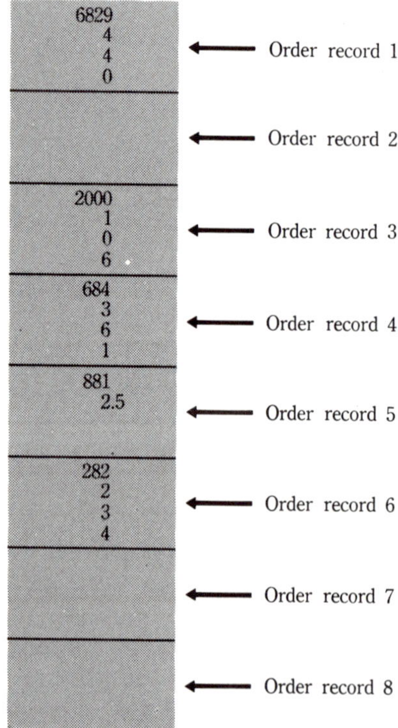

FIG. 4-6. 225 TABSOL MEMORY LAYOUT

cessor fields will contain the record numbers of the successor and predecessor orders. Thus, the QUEUE RECORD should tell us that ORDER RECORD 3 contains information about the first order in queue (in other words FIRSTO should equal 3), because this order has the lowest priority number. The successor of ORDER RECORD 3 is ORDER RECORD 6, and so forth. LASTO should equal 1 because ORDER RECORD 1 contains information about the last order in queue.

Assume that the record number of the order to be filed (Recond number 5) has been noted in the non-record parameter RECNUM, and also assume that the priority of the order to be filed (priority 2.5) has been noted in a non-record parameter PRI. We begin by getting the record for this queue in order to determine if there is a first order in the queue.

TABLE 1

RECORD	GO TO
GET QUEUE	TABLE 2

If there isn't, go to the table that handles such situations: namely, TABLE EMPTY.

TABLE 2

FIRSTO	ORDER	RECORD	GO TO
GR 000000	PUT FIRSTO	GET ORDER	TABLE 3
			TABLE EMPTY *

Compare the priority PRI of the new order with the priority PRIOR of the order already in queue.

TABLE 3

PRI	SUCCES	PREDEC	ORDER	RECORD	GO TO
GR PRIOR	GR 000000	~	PUT SUCCES	GET SUCCES	TABLE 3
GR PRIOR	EQ 000000	~	~	~	TABLE LAST *
NGR PRIOR	~	GR 000000	~	~	TABLE 4
NGR PRIOR	~	EQ 000000	~	~	TABLE FIRST *

* Not shown.

Let's pause for a moment and think about the four possibilities. In row 1 we have the case where the new order has a larger priority (hence a worse priority) than the order in queue. The order in queue does have another order behind it so that we can get the successor record and try again. In row 2 the new order has a worse priority than the order in queue, but the end of the queue has been reached so TABLE LAST is called into action. In row 3, the new order has at least as good a priority as the order in queue, and the order in queue is not first because it has a non-zero predecessor. This is the typical situation and TABLE 4 will shortly handle it. Finally, in row 4, the new order has at least as good a priority as the order in queue, but this happens to be the first order in queue, so TABLE FIRST gets the call.

TABLE 4

PREPRE	PREDEC	RECORD	PREDEC
PUT PREDEC	PUT RECNUM	GET PREDEC	PUT PREPRE
SUCCES	RECORD	SUCCES	GO TO
PUT ORDER	GET PREDEC	PUT RECNUM	TABLE DONE

The current predecessor of the order that has just been beaten is noted in the non-record parameter PREPRE (previous predecessor). Then the new order RECNUM is made the new predecessor. Then the new order record, RECNUM, is gotten and PREPRE becomes its predecessor, and ORDER which has faithfully saved the record number of the order that was beaten becomes the successor of the new order. Finally, the order record for PREPRE is gotten and RECNUM is made its successor.

Conclusion

The Company's various experimental tabular languages, such as the 225 TABSOL just described, have proven remarkably adequate. However, the added power of an algorithmic language seemed very appealing, particularly as the prospects for structure table application in all kinds of problem areas brightened. As a result, General Electric's

Computer Department has incorporated structure tables under the trade name TABSOL as an integral part of GECOM—a COBOL-based General Compiler for the GE 225 and 215 computers. Joining TABSOL with GECOM places the power of a full-fledged algorithmic language at the command of every structure table block. Thus, while structure tables might not be the answer to all programming problems, the applicability of these tabular approaches has now been considerably enhanced.

This article, by T. F. KAVANAGH and M. ALLEN, originally appeared under the title "The Use of Decision Tables" in *Data Processing, Vol. VI* (the Proceedings of the 1963 International Data Processing Conference). It is reprinted here with permission of the Data Processing Management Association, Park Ridge, Ill.

5

Decision Tables in the 1964 Census of Agriculture: Three Views

View 1

by Thomas Breen, who was responsible for producing the specifications for the Census of Agriculture (1964).

My approach to this discussion of decision tables is strictly subjective. I am not attempting to express or even to recommend dogma. I am stating personal experience, opinion, and suggestions. Approximately twelve months ago, I was vaguely aware of the term "Decision Table" and knew that the Agriculture Division was planning to use this format to express specifications for the computer editing and coding of the individual records and to some extent for the tabulations required.

Some 600 pages of decision tables later, I feel somewhat as if I were better acquainted with the subject. As of now I like them, in fact, if I were the missionary type, I'd be quite zealous in expounding their use as gospel. I consider them to be a powerful tool in providing a means for requiring precise, unambiguous specifications. There is no guarantee that decision tables per se will automatically provide such specifications. They can provide them and the tables can readily be examined to see that they are precise and unambiguous. The subject matter division is enabled to evaluate the completeness and logic of the specifications it establishes. As may be pointed out by other members of the panel this can be done with mathematical exactness under many circumstances.

I am not in a position to comment on the use of decision tables in the multitudinous housekeeping, control and other aspects associated with the comprehensive computer program. I have been exposed over the several years to courses for programming Univac I, the Univac 1105, and the Univac 1107; but have not had any practical experience in programming, not even to the extent of utilizing FORTRAN for a special test or the like.

In previous Agriculture Censuses, the clerical edit and coding instructions and the mechanical edit specifications were largely expressed in narrative form; they were subject to revision as the actual work progressed and included a significant proportion of "refer to the technical staff." Although this later instruction is still available in the form of "print out for review," our approach has been to provide an editing or imputation procedure which in all instances, even those calling for "print out," will retain the record with consistent data for tabulation, although it may be changed by a later correction pass if it is determined by review that changes were not handled properly.

The preparation of comparable narrative specifications for computer programming was not even considered seriously by the Agriculture Division. The potential sources of errors and misinterpretations in inexplicit statements were frightening. Consider only differentiating among wheat, winter wheat, spring wheat, durum wheat, and other spring wheat, all of which appear as separate items for one or more of the different state varieties of the basic Agriculture questionnaire. The actual volume of narrative needed hasn't been estimated but would be substantially more than the equivalent specification in the decision table format.

To express the various relationships inherent in the some 355 questions possible on the questionnaire, to examine the inconsistencies possible from enumeration errors, clerical processing errors (although clerical processing is limited it is still required for a few items; coding write-in crops or the like), and card punching errors (this is our first large scale experience in string punching) has been an overwhelming job. It has been accomplished to a large extent in about ten months with a group which at no time exceeded five professional staff members and one secretary-typist. Just how well it has been done remains to be seen but at least it can be seen. The tables exist; relationships tested can be traced to the specific tables and steps involved. We have looked at the results of some of the preliminary tests. We've seen where the data did not get handled the way we thought we wanted to have them handled. A recheck to the tables showed that in at least one instance we had neglected a step: in other instances the program was not following the actions indicated by the tables. In such cases the programmer could review the actual step by step tests and see that an error had occurred and hopefully know the location in the actual program where it occurred.

Obviously, we've had some help, largely from Mr. O'Brien who spent several months educating us on the potentialities and possibilities of decision tables, on the tests of completeness and independence and other related information while we in turn imparted fascinating information on the average yield of potatoes, the expected proportions of lambs and ewes in a typical flock and the average price of cattle sold. The three other members of the staff participating in the preparation of the tables had had no previous experience with their use. Yet in a comparatively short time, three to four weeks as an average, they were preparing tables involving the various sections of the questionnaire using only some very general instructions as to the action desired.

We do not anticipate any significant problem in having the decision tables programmed to provide a satisfactory computer output. Certainly there will be some mistakes, logical and otherwise, discovered and corrected in debugging the program. Probably not all the mistakes will be found. Analysis of the forthcoming tables will provide a test of how well the job has been done. Since some 6–8 hundred items will be tabulated for each of 3,100 counties, the review will be a substantial task. Members of our professional staff for the most part are not now acquainted with the precise details of the decision tables. We are preparing to give them a narrative statement of the general actions intended to be accomplished by the tables, a detailed explanation of one or two more important groups of the tables, and a complete set of the detailed tables. It will be interesting to learn how easy or difficult it will be for them to comprehend these procedures for use in review of the tabulations.

I believe there is much to be gained from the use of decision tables and, at least for my part, a great deal more to be learned about how to use them most efficiently and most effectively. There should be some gain whenever a program involves a large number of records, if the order of testing is arranged so that the situations occurring most frequently could be acted upon with the minimum number of tests. That is, the frequency of occurrence should have some relationship to the order of testing. Some attempt was made to follow this logic in preparing our tables; however in many instances we had only limited information on which to base our estimates of frequency. We have made provisions for obtaining some counts in our diary output which should prove useful in subsequent planning and preparation of decision tables for a future Agriculture Census.

There are obviously some areas in which the programmer or systems analyst should be of considerable assistance to the preparer of the decision tables. As a case in point some 75 pages of tables relating to use of fertilizer contained some highly repetitive tests and actions with variations for certain crops and different states. I understand the programmer after an analysis of the tables was able to specify a relatively limited number of steps or instructions for the computer with variations dependent on the crop or state code. Perhaps these steps could have been incorporated in the tables or perhaps they should have been left to the programmer's analysis. I think the proper approach will best utilize the knowledge and skills of both the subject matter and the computer technicians to arrive at efficient programs.

To introduce a slightly different but nonetheless related subject, I believe that some "recommended conventions" could be developed to lessen potential misunderstandings and logical errors. As an example the following statements

$$X = O \qquad Y \mid N$$

and

$$X \neq O \qquad N \mid Y$$

are identical. Unfortunately we used both statements at random in our tables and were occasionally confused by the double negative in the second test. We could easily have avoided its use. There are potentials for misunderstandings or errors in the $\leq$ and $\geq$ relations if the table maker is not careful. I suggest we prepare some training material for the subject matter analyst to use in learning how to prepare and use decision tables, and that we include the "recommended conventions" in such material. I am not concerned about the ability of a pre-processor or compiler to recognize the conventions as much as I am about the case of misunderstanding for the person who is reviewing the decision tables prepared in the subject matter division.

View 2

by BROADUS H. KING, JR., who supervised the preparation of the computer programs for the Census of Agriculture in 1964. The specifications for the edit and imputation procedures were presented in decision table format.

As a little bit of reverse twist I would like to begin my remarks by stating my general conclusion on the value of decision tables as a specification medium for the programming effort for the Census of Agriculture.

"The very extensive and comprehensive consistency checks and resulting adjustments in data considered desirable and profitable by the Agriculture Division could NOT have been computerized for the 1964 Census without the use of decision tables."

To understand how I reached this conclusion one must know some of the conditions existing in the environment in which the computer programs were to be prepared.

An extremely tight production schedule for producing the processing programs was automatically created. The enumeration was to take place in November 1964, and the first reports had to be published by March 31, 1965. As of January 1, 1964, no detailed edit and imputation specifications had been forwarded to Demographic Operations Division (DOD).

There were no computerized specifications for edit and imputation which could be used as base for the specifications required at this time. In previous Censuses it was necessary to record various parts of the questionnaire on different punch cards and the relationships existing between the parts could not be examined mechanically. In the computerized process the questionnaire could be examined as an entity and presumable higher quality data would result.

Knowledge of Agriculture subject matter in the DOD programming staff was extremely limited. Interpretation of any type of generalized specifications, particularly narrative, would have been very lengthy and error prone. There was neither time nor professional resources available to use the continuing conference technique whereby programmers become more or less indoctrinated in subject matter.

Certain characteristics of the application itself are also pertinent: The Census of Agriculture questionnaire has approximately 1,500 data fields for which edit and imputation specifications had to be provided. Since any part or parts of a schedule might contain data for any one farm and since definable relationships exist between the various parts it was most desirable to treat this record as an entity rather than subdivide it to a number of smaller records. Subsequent experience has shown that the decision tables expressing these relationships required approximately 600 pages and resulted in 20-25,000 lines of coding.

Considering all of the above it was decided that decision tables, though untried on a large project of this nature, seemed to provide the most practical way of communicating the edit specifications in the time frame allotted. Thus the project was undertaken using the decision table approach. I will describe the experiences of DOD in working with the tables in terms of advantages and problem area.

Technically the decision tables were very effective as a medium for communicating the detailed logic required in the edit and imputation procedures. Other techniques with which the decision tables can be compared are narrative descriptions, oral specifications, or flowcharts. As soon as one becomes aware of the amount and complexity of the detail involved in the Agriculture edit it becomes almost self-evident that narrative and/or oral specifications offer a costly, impractical solution to the problem. A concise format which could readily present complete detail was required and decision tables fit the bill. In practice we found the following advantages in using the decision tables:

1. Increased productivity of the programming staff and compressed over-all length of the production effort. The need for interpretive conferences with the Agriculture Division was minimized. Planners could turn the work over to programmers in a much shorter period of time and with less effort on their part. Although we have not yet run the production which will offer the final proof, it appears that misinterpretation of the specifications will be insignificant.
2. The technique forced the subject matter people to discover, resolve, and document the solution to numerous detailed problems that would ordinarily come to light in the middle of the programming effort or even after production started.
3. The decision tables aided production by providing an excellent basis for production control. It was easier to make a more accurate estimate of the portion of the total specifications received. Even more important the decision tables provided an excellent basis for estimating the programming effort required to complete that phase of the job.

4. In some of the more straightforward cases the table could be converted to coding almost on a line for line basis. These tables provided excellent experience for trainee programmers.

At this point, to attempt to prevent extrapolation of my remarks as meaning more than is intended, I want to emphasize programming functions not eliminated by the decision tables.

1. System design—run structures, facility utilization and processing techniques.
2. Program design
   a. develop file structures
   b. design record layouts
   c. technique of addressing record
   d. parameter table structures and addressing techniques
   e. code structures
   f. housekeeping—clearing of work areas at appropriate control breaks, processing flow, initiation, close out, etc.
3. Data control—error detection, notification, and recovery procedures (label checks, sequence checks, record counts, etc.)

The advantages previously mentioned were attained even though certain problems were encountered. There were some difficulties with compatibility between subject matter techniques for accomplishing a desired objective and the structure of the computerized system or programs. One example out of several hundred where necessary changes were uncovered involved the method of determining irrigated acres for a particular crop. The logic expressed in the decision tables was correct, but a direct translation of the technique gave incorrect analysis figures in the diary and could have resulted in erroneous data being inserted in the detail record in the correction pass. The tables had to be modified. To detect as many of this type situation as possible, DOD found it desirable to have the tables reviewed by a professional knowledgeable in subject matter, system structures, and programming,

Another area of some concern was the structure of the tables themselves, particularly in regard to the amount of detail required in iterative type situations. On one hand every iteration was recorded separately if there were even the most minor variation in procedure (e.g., selection of a different parameter). This resulted in an inordinate amount of paper being generated. At the other extreme logic was expressed in iterative form (e.g., "Do for all crops"), but procedures were not carried to the lowest level of detail. This allowed very concise statements of the specifications but opened the door to possible oversights by the programmers. For example in one such set of tables the procedures did not consider the fact that various units of measure involved were scaled differently, and therefore required different rounding procedures in arithmetic operations. A programmer not acquainted with the subject matter would not have solved this problem in his coding. This was another type of situation where DOD found the review of the tables by a professional knowing both subject matter and programming to be very profitable.

The logistic aspect of correcting and maintaining the documented tables appeared to some to be burdensome. However, I believe this conclusion is chiefly the result of a comparison of this system with one using an absolute minimum of documentation. I feel that decision tables are just as easy or easier to maintain than any other type of detailed documentation.

Perhaps the most serious problem encountered in the use of the decision tables was a psychological one rather than a technical one. Experienced programmers used to working under a system where extensive subject matter knowledge on the part of the programmer was substituted for detailed documented specifications found themselves unable to adjust to decision tables. They felt subject matter knowledge an essential ingredient in their professional stature, and to them decision tables were a blow to their pride. They felt they were being relegated to the role of transcribing clerks. Discussions emphasizing the need for senior personnel to perform the planning functions previously described were ineffectual. In some cases the only solution seemed to be to assign these personnel to parts of the project not directly associated with decision tables.

View 3

by James L. O'Brien, who worked on edit specifications for the Census of Agriculture (1964).

At the time it was decided to prepare specifications for the 1964 Agricultural Census in Decision Table form, actual Bureau programming experience with this format consisted of several limited-scale programs turned out by the Program Research Branch in Data Processing Specialty Division, and the Foreign Trade Export Program, where the tables were prepared by DPSD programmers and reviewed by a task force which included all interested divisions.

The Agriculture Census represented a full-scale test of a different type than the Foreign Trade Program, and covered the preparation of editing specifications for the first computer edit of a complicated questionnaire, with a minimum of 500 data fields and a fifty-state total of 1,500 involving some 3,500,000 returns, a total of 53 man-years of programming with 14 man-years on the edit program, and about 3,000 hours of 1107 computer time. These edit specifications comprise about 600 pages of decision tables, and cover practically all data items and thousands of individual relationships. The computer performs a complete edit; no questionnaire is rejected for manual correction, although there is provision for such correction.

The specifications were written by a single experienced subject-matter analyst, the Assistant Chief of the Agriculture Division, working with a staff of one to four persons brought in from outside the division. In preparing and reviewing specifications in decision table format, the specialist was able to see in detail the logical consequences of proposed tests and to determine the appropriate actions on the spot. He learned all he needed to know about decision tables without difficulty, prepared hundreds himself, and reviewed all of them. On the other hand, persons not specifically familiar with Agriculture Censuses (like the writer) or decision tables (like other members of the staff) were able to turn out completed tables from informal oral specifications, formulating questions on appropriate tests and actions in explicit problem-oriented terms as needed.

The advantages, problems, and promise of decision tables in the programming process itself are treated in other papers and will not be discussed here. Nevertheless, as one who has had to "eat his own cooking" by working with programmers in the Demographic Operations Division on tables he prepared in the Agriculture Division, the writer believes that the following problem areas and potential problem areas should be indicated.

A. NEED FOR SUPPLEMENTAL EXPLANATORY MATERIAL

At one time it was felt that decision tables might obviate the need for other types of edit specifications. It now seems clear that some form of supplementary material is highly desirable, at least in extensive and complicated edits. The most efficient sequence of decision table testing is often not the easiest pattern to understand, and the precise rationale behind specified tests and actions may be difficult to discern—and months later can become hazy to the originator himself—without some description of the why behind the what.

On the basis of experience to date, the preparation of such supplementary explanations does not appear overly burdensome, especially if it is done by the originator of a set of tables at the time when the tables are written.

B. TEMPTATION FOR ANALYST TO EXTEND TESTING AND ENGAGE IN PROGRAMMING

Anyone who has written a complicated decision table and then tried to describe the tests and actions fully in narrative form, without reference to the table, realizes how difficult it can be to keep the narrative unambiguous and still intelligible (hence the wide use of flow charts by programmers). Once some practice has been gained with decision tables, it is relatively easy to add tests until a table provides elaborate cures for which there is no disease—more technically, a disproportionate programming effort may be exerted to deal with very rare and harmless cases. In addition, lack of detailed knowledge of computer characteristics and program structure may lead the analyst to incorporate in decision tables procedures that are less efficient than a skilled programmer could devise.

C. DESIRABILITY OF INDEPENDENT REVIEW

The number of logical and typographical mistakes in six hundred pages of decision tables is likely to be significant despite the most painstaking scrutiny by the originating group. The mistakes that have been discovered even after such checking argue strongly for a critical independent review where resources are available. Considering the time lag before the Agriculture Edit Program will be tested on the computer, such review, especially by systems personnel familiar with both programming and agriculture, has been particularly worthwhile and has probably cut down markedly the time and effort that will be required for logical debugging and program correction.

D. NEED FOR A RAPID, SIMPLIFIED REVISION SYSTEM

Besides the discovery of mistakes and the need to provide clarifying notes on tables, changes resulted from revisions in subject matter treatment of certain crops, states, and situations. Whereas with narrative specifications these revisions would be issued in the form of a few general notices affecting a great many programs, with decision tables, individual tables, sometimes in sizable numbers, had to be changed. Although this feature permits the analyst to determine realistically the effect of a proposed change, it results in a great deal of paper shuffling. When there are many changes, and especially if a large number of copies of decision tables is issued, keeping up with revisions became a serious burden for the subject matter division and a vexing problem to systems and programming personnel, a situation which over time is likely to have the pernicious effect of discouraging desirable change. One partial solution is to limit issuance of revisions to a rigidly restricted list of recipients (say three to five) who actually need them, and make it clear that only these copies will be kept completely up-to-date.

Although there is general agreement on many aspects of decision tables, there are still several areas where additional study, experience, and perhaps even experiment are desirable. Some questions raised by the Agriculture Census experience are:

a. What is the most effective level of detail to include in decision tables so that they remain problem-oriented and still provide all the information needed by their most important user, a machine-oriented programmer, without freezing the program into an inefficient pattern?
b. What is the optimal degree and pattern of participation by programmers in the preparation of specifications in decision table form? It seems evident that more active participation, at least in the early stages of the process, would be beneficial.
c. Which types of supplemental interpretive and explanatory material are most effective? Further exploration of this area, with some measure of differences in learning time or programming time, would seem worthwhile.
d. Which changes in the roles and training of analysts and programmers, and in decision table format and structure will be required to take fullest advantage of the CENTAB translator, which will produce coded programs directly from the tables themselves?

e. Pending an operational CENTAB translator, what could profitably be done to adapt FORTRAN to produce test programs from decision table specifications, and possibly to generate test data so that there will be quick feedback to the analyst for logical debugging?

f. What measures are necessary to insure that the chain of documentation from general explanation of editing objectives to final computer output is readily comprehensible to persons familiar with the appropriate discipline (subject matter, programming, sampling, or whatever) but new to the project? Decision tables appear to be one useful link in this chain—several commercial concerns believe that their main contribution is precisely in improving documentation. As a step in strengthening documentation—an area which many Census Bureau personnel consider to be weak—and reducing excessive dependence on individual expertise, perhaps an independent team audit of the documentation chain would be justified.

g. What is the feasibility of extending computer editing to permit selection among alternative programs on the basis of analyzing the effects of the editing programs themselves on early returns? Such arrangement might be useful where tests were included with no knowledge of the character and frequency of expected errors, and early results indicate that isolatable blocks of tests and actions are not affecting the statistics.

h. How can we more closely relate the economic value of improvements in published data to costs resulting from separable blocks of editing specifications? One approach might be to analyze some cost-benefit ratio of a given table or group of tables before they are released. Costs of programming most decision tables can now be determined quite closely; costs of testing and running, and expected frequencies of errors can be estimated, actual costs and the actual impact on published data can be accurately determined *after the fact* with existing decision table and computer techniques.

The appearance of serious predictable errors not corrected by an edit program points up dramatically the corrective steps that might have been taken; there is no such indicator to warn us of edit specifications and programs in operation that cost more than they are contributing. To paraphrase a well known poem,

> The saddest words of tongue or pen
> Perhaps are not *"It might have been,"*
> They might be these we daily see,
> "It is, but hadn't ought to be."

In summary, before the Foreign Trade experience, it was known that decision tables offered promise as a useful and powerful device for efficient logical and statistical manipulation of relationships which interact in complicated ways. The Foreign Trade experience also showed that they were reasonably efficient even without manipulation and could be understood by persons who were familiar with the subject and with previous computer procedures. With the Agriculture Census, despite numerous problems, no critical shortcomings of decision tables were uncovered, and it now seems clear that they represent a practical unambiguous method of transmitting large numbers of complex specifications which might be of value elsewhere in the Bureau.

The foregoing material was first printed in *Seminar on Decision Tables: Two Years of Experience at the U.S. Bureau of Census*, Washington, D.C., September 30, 1964.

6

Decision Tables at the Bureau of the Census

Two years ago I had the good fortune to attend a symposium on decision tables in New York and I came away impressed enough to write, in a report on the symposium, that, with respect to improving programmer productivity, "there is nothing in sight quite so prospectively powerful as decision tables." At about the same time Mr. Solomon had reached a similar conclusion from literature he had seen and we early joined forces to push the development and use of decision tables through the Bureau. Since that time I personally have prepared thousands of decision tables and my enthusiasm for them has increased over the years. The spread of the use of decision tables throughout the Bureau has not matched my enthusiasm, however. This has led me to speculate much about the programming process and to develop an interpretation of programming development at the Bureau over the past few years which I hope will make some sense out of what has happened and offer some rational look into the future.

Two years ago, programming presented a particularly acute problem at the Bureau. Large numbers of programmers were required for closing out the 1960 Census and in preparing for the Economic Censuses. The Agriculture Census was in the offing and a new computer system, the 1107, was arriving, both appearing to require massive programming support. Some of us looking at the 1107 instruction repertoire were dismayed with its complexity and feared it would become another costly drain upon programmer talent. At the same time, sponsors throughout the Bureau were raising complaints that they could not get quick programming service for the many small but pressing jobs coming in and that programming costs for such jobs were so unreasonably high that the Bureau was effectively priced out of the performance of important services. Furthermore, they complained that research in the Bureau was being stifled from lack of easy and economical access to the computers by their analysts. Finally, administrators were becoming alarmed at the heavy dependence important operations in the Bureau owed to a few key programmers who carried most of the specifications and know-how for the operation around in their head. Thus in this atmosphere some of us concluded that the major need was some means of improving communication between programmers and computer and between analyst and programmer, the first to ease the coding and debugging chore and the second to relieve the programmer of some of the burden of recasting subject matter specifications into computer specifications.

About this time I grasped at 1105 COBOL and found encouraging prospects there for better documentation and ease of programming. However, I became somewhat disenchanted when, several months after I had written a fairly complex COBOL program in the usual fashion, planning in my head as I coded, I tried to make a minor modification. The more I tried the more confused the result and finally I had to discard the program and rewrite it from scratch. With this experience and in this atmosphere, then, I attended the Decision Table Symposium. Thus it was not surprising that I should then have seen so much promise in decision tables.

It early developed that decision tables fell into two classes of application, one at the analytical level and the other at the programming level. At the analytical level, decision logic tables displayed their power specifically in the area of data edit specifications where complex interrelations of consistency among many items in a data record had to be displayed and understood before the edit could be programmed easily and controlled effectively. The first major application in this area was in organizing the Export data edit for the 1107. Some of us directly involved were pleased at how easily the edit was understood and accepted by the supervising committee once it was cast in decision logic form. Some of the analysts concerned were most grateful to be able for the first time to understand the edit in detail and as a whole, and I believe Mr. Kahn will testify that the analysts were able to provide much assistance in the debugging process because they could follow what was going on. In this first application the decision logic tables were reviewed by subject matter analysts but not prepared by them. Nonetheless, it was demonstrated that a programming staff with no experience in the specific subject matter could organize a job rather easily where clear communication could be established with subject matter analysts by means of decision logic tables.

About this same time there was some experimentation with having Population Division analysts prepare decision logic tables for tabulations of files such as the 1960 Census 1-1000 sample of population. Here, however, it was soon discovered that the bulk of specifications for a tabulation were simple enough to be presented to the computer by the analysts as a list of parameters. A tabulation generator, written in FORTRAN and organized by decision tables, was soon prepared to accept such a list, and henceforth decision logic tables were prepared by Population Division analysts only where specifications for a particular table parameter were complex.

Also, about this time, an unrelated development was taking place that contributed more directly toward closing the gap between analyst and computer. A 1401 IBM computer with a FORTRAN package had been installed at the Bureau, and this, with the early prospect of a FORTRAN compiler for the 1107, encouraged many analysts to undertake their own programming of their research and small job applications. This development did not involve decision tables, mainly because the applications were so narrowly specialized that analysts could organize them successfully enough in their heads and with the FORTRAN language itself. Of late, however, some analysts have been overreaching their capacity to remember and their ability to organize increasingly complex data processing jobs as they code. Consequently, some have been turning to decision logic tables.

The first major application where analysts themselves have prepared large numbers of decision tables has been the Agriculture Census edit. Some of those involved will testify that decision tables not only permitted quick and accurate organizing of a stupendous mass of detail but ensured communication with programmers in a manner that minimized the amount of effort required to translate the specifications to computer coding.

There also have been recent increases in the use of decision tables by analysts to work out the implications of specifications such as those of the import edit for Foreign Trade. In addition. support has come from analysts for a decision table processor that would permit more or less direct translation of decision tables to computer programs. Finally, analysts are encouraging the development of generators for tabulation, for statistical analysis, for report preparation, and for graphics.

Thus, over the last two years, there has been much improvement in increasing the access of analysts to computers and in utilizing analysts to decrease the load upon programmers for translation of subject matter specifications to coding. Decision tables have contributed heavily to this development.

At the programming level, however, the spread of the use of decision tables has been most disappointing. In fact, I know of no intensive use of decision tables as a direct programming aid outside my own Branch. By direct programming aid I mean the use of decision logic tables to organize the detailed coding required to direct a computer. The main explanation I can offer is that the pressures upon programming resources that were anticipated two years ago have not materialized and that the pressing need for sharp increases in programmer productivity has disappeared. Several factors appear to be responsible. One is that the delivery of operational hardware and software for the 1107 has been a year later than some of us had expected. Thus, intensive conversion to the 1107 has not yet begun and, in any event, now will occur when programming activity for the Economic and Agriculture Censuses has eased somewhat. Another factor has been the spread of the use of FORTRAN among analysts in the Bureau. Practically every division now has analysts who can get small jobs done expeditiously with FORTRAN on the 1401 and on the 1107. Two to three hours of 1107 computer time per day are now required for this do-it-yourself application. Thus, pressure on programming resources for research and small job work has largely disappeared, and a heavy pressure this was, since much was mathematical and logically involved, presenting unusual difficulties for the standard programming techniques.

It would appear, now that programming resources apparently are ample, we ought no longer be concerned with drastic measures for changing programming techniques. The time tested practice of assigning a programmer to a particular task for life, so as to speak, letting him acquire over time a personal and private bag of tools for getting a specific job done as efficiently as possible on a specific computer, cannot be improved upon as long as the job, the programmer and the computer do not change. This practice, however, is most insidious because it is so attractive, to the analyst who can avoid thinking about a job, to the administrator who can avoid knowing what it takes to get a job done, and to the programmer who can avoid revealing for review how he gets a job done. Thus, it is no wonder that the loss of a key programmer is a catastrophe, that revisions to a job are strenuously resisted, that the first cry when a new computer arrives is for a device to permit the old programs to run on the new computer. It is not farfetched to suggest that the present crisis in operating the 1107 arises from applying to the 1107, programming methods effective on the 1105. The 1107 simply shows up the primitive character of much of our programming techniques. Traditional practices are most vulnerable to change. This is reflected in the attitudes of programmers themselves. Some pessimists among them hold that we have already reached the point of no return, that we have neither the resources or the intellect nor indeed the economic need to convert to new programming techniques and new computer systems. Old programmers in moments of despair will tell you that a job that cost ten dollars on Univac I, costs one hundred dollars on the 1105 and will cost one thousand dollars on the 1107.

Our experience has been to the contrary. Over the past two years we have tested a decision table and FORTRAN or COBOL combination in hundreds of jobs. In every case we have been astounded with the gains in programmer productivity and scope. As I have said many times in the last two years, with this combination I have felt for the first time in my programming career that there is literally no data processing job that cannot be undertaken with confidence and ease. The power of decision logic tables as a tool for organizing the most complex of data processing and control operations is phenomenal. In addition, FORTRAN and COBOL are "natural" languages to use in writing decision tables in the sense that they express operations economically and still permit automatic coding from decision tables. In a strict sense we program with decision tables in the old manner of planning as we code, and coding, in the technical sense, becomes merely a transcription from the decision tables. . . . This mode of organizing a job is fast and accurate and changes can be incorporated easily. In addition, the logic of the job is debugged from these tables, not from coding or memory dumps. With this well thought out and compact control over a job, a programmer's productivity becomes phenomenally high.

Our range of application of decision tables over the past two years has been wide. With the decision tables, 1105 COBOL combination, we undertook dozens of research projects involving the Foreign Trade export edit, analyzing rejects by cause and frequency, tabulating cell density by sample—non-sample card count—thus contributing a foundation for some of the edit specifications. In addition, with this combination we easily developed a program providing a complicated comparison of the outputs of the old and new edits, applying some procedures an analyst would apply in deciding whether a discrepancy was serious, thus easing and speeding the analyst's review. This program proved very valuable in detecting errors in specification and helped account for the early and easy debugging of the edit. Another program we undertook with this combination was a display of the actual edit paths taken by the detail records. These programs could not have been produced with the conventional programming talent then available. Incidentally, both these programs, originally written in 1105 COBOL, were translated to 1107 COBOL and successfully run on the 1107. These applications convinced me of the ease with which the programming of new areas of analytical review and control can be undertaken.

As mentioned previously, a FORTRAN tally generator was written and has been run successfully on at least fifty jobs. This generator would not have been possible if it had not been organized by means of decision tables. This job was done at one third the cost estimated for conventional techniques, and this despite the fact that the method of handling the job and the coding generated resulted in the job consuming perhaps four times more central computer time that would have been taken by a conventionally coded program. If we assume that the program was produced with one-fifth of the effort required by conventional programming, the implication is that for the total cost to be cut by one third, the ratio of programmer to total costs in the original estimate was about 97%, which seems not unreasonable for this particular job. The success of this generator has convinced me of the ease with which generators in other areas can be put together. This application to me appears to offer incredible prospects for increases in programmer productivity and in access of the analyst to command and control of the computer.

Some prices may have to be paid, however. New areas of experimentation cost time and money. New applications may not be economical. In fact our present Decision Table–FORTRAN combination is so independent of the computer that particular features of the computer resulting in economical coding are not taken advantage of. Thus programs compiled from this combination require perhaps 4 times the central computer production time that more machine-oriented programs would require. We probably cannot afford such programs unless the ratio of programming to total costs exceeds perhaps 70%. In any event, small jobs and rush jobs can now be undertaken with ease.

If we enumerate the prospects that decision tables open up before us, however, they are considerable. The first is that the most complex jobs can now be undertaken with confidence. Decision tables permit the breaking up of a job in small pieces, the separate study, understanding, development, and documentation of each, and the combination into a workable whole. We have used decision tables to organize a set of decision tables. The second stems from the first.

We can now build the most complicated generators with ease and thus offer the analyst a means of addressing the computer in his own language and offer the programmer a means of handling with ease the incredible amount of detail required for a job. The third is that decision tables for the first time provide a means of studying, systematizing, and documenting the intricacies of data processing logic, placing the training in, and the development of, data processing on an explicit basis. The fourth is that the maintenance and change of a program can be transferred at will to new staffs, introducing variety and equalizing loads on programming staffs. The fifth is that a reorganization of the programming staff is possible along new specialities, a planning and documentation staff independent of subject matter, language and computer, and a staff specializing in languages and computers. With all these prospects we should approach the day when new concepts required to get new jobs done or old jobs done in new ways can become almost automatically translated to computer coding. Then will computers become natural extensions of the intellect, and change no longer a cause for despair.

This address, by RICHARD A. HORNSETH, was given at the *Seminar on Decision Tables; Two Years of Experience at the U.S. Bureau of Census*, Washington, D.C., September 30, 1964.

7

Planning Networks and Resource Allocation

The arrowed diagram has become well established as a basis for time planning, but it is less widely used for the detailed allocation of resources. It is generally considered that the factors involved in specifying resource requirements and the conditions to be satisfied when allocating resources, vary so much from project to project that any generalized computer approach is impracticable. Early network-based resource allocation programs had severe restrictions in the manner by which activity resource requirements and project resource availabilities could be expressed and thus their usage was inhibited.

The basic problem appears simple enough at first sight and is usually to calculate either:

a. The minimum project duration if only fixed levels of resources are available; or
b. The minimum quantity of resources (and their most efficient development) if the end date of the project is considered to be fixed.

These solutions are obtained by scheduling the resource requirements for individual network activities against resource availabilities and deploying activity float to obtain the best result. A typical example of such an operation on a simple network is shown in Fig. 7-1 (a) and (b).

However, real projects are much more complex than this example. The network may comprise several thousand activities and involve more than a hundred different types of resources. Also, in practice many special scheduling conditions have to be considered when formulating work schedules. Some activities involve the use of several resources at different points of time; some must be worked continuously once started

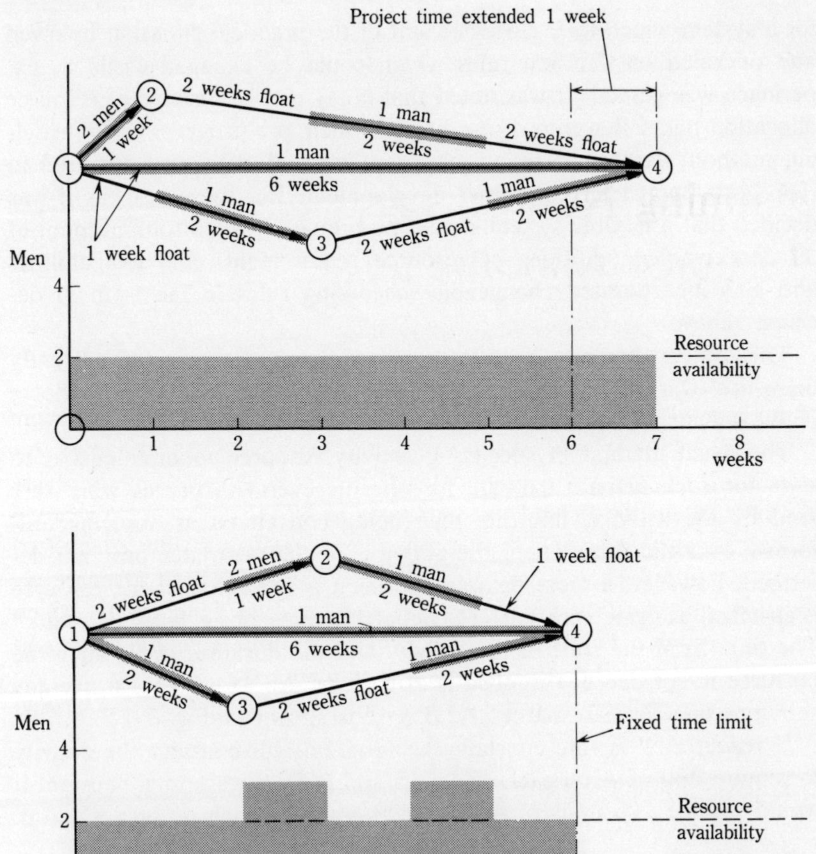

FIG. 7-1. RESOURCE ALLOCATION: (A) ALLOCATION WITHIN RESOURCE LIMIT; (B) ALLOCATION WITHIN TIME LIMIT

and others may be stopped if necessary, etc. It is the combination of size and complexity which has restricted the development of computer-based resource allocation models in the past.

A few years ago it was realized in Britain that there were substantial benefits to be gained from the effective deployment of computers to the allocation of resources on building sites and in factories, etc., and a determined onslaught was made on the practical problems which had so far restricted progress. It was clear from the outset that the need was for a system which took close account of the practical situation involved and operated on decision rules which could be changed easily as experience was gained. It was noted that many earlier systems of resource allocation had fallen into disuse because their pre-programmed scheduling method was found to be inappropriate and the effort required in reprogramming deterred further development. For these reasons it was decided that a flexible system should be prepared which took account of all conceivable conditions of resource requirements and availabilities and also incorporated changeable scheduling rules in the form of decision tables.

This system[1] has now been in operation for over a year and currently there are over 100 users of the program in Britain and Europe. Some of the salient features are now described.

The usual method of specifying activity resource requirements is to state for each activity the rate for one of several resources which are used by the activity, this rate then being considered as applying uniformly over the duration of the activity. In the computer program described, however, a more flexible approach is used. Either the resource is specified as "rate constant" (as defined above) or as "total constant." The time from the start of the activity and the duration over which the resource is (or can be) applied is also given. An example of a five-day activity using three resources A, B & C is shown in Fig. 7-2.

If resource A is rate constant the program will consider the activity to require 200 of A on each of days 5 and 6. If A were total constant it would assume 100 units of A (i.e., 200 ÷ 2) on each of days 5 and 6.

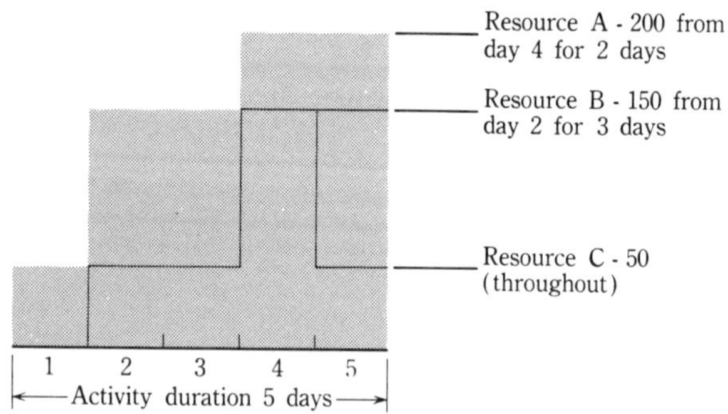

FIG. 7-2. ACTIVITY WITH VARIABLE RATE OF RESOURCE USAGE

[1] CT 1900 Series PERT (International Computers & Tabulators Ltd. London 1966.)

Similarly, if B is rate constant the assumed usage is 150 units for each of days 2, 3 and 4. If resource C is rate constant it specifies 50 units of resource for each day of the activity duration and if total constant it would be 10 units (i.e. 50 ÷ activity duration) for each day. This not only gives a comprehensive method of specifying variable rates of resource requirements but also provides a means of minimizing alterations to input data when it is necessary to adjust activity durations. For example, when increasing the duration of an activity using multiple resources, some resources will increase proportionally to the activity duration (these can be specified as rate constant) but others will not increase in total (these can be specified as total constant). A special designation in the program enables activity-terminating resources to be specified from the end of activity duration, thus further simplifying modifications.

Work Continuity

In order to completely specify the characteristics of the work to be executed, a further set of data is necessary in order to indicate the required coninuity of the work and any obligatory association with preceding and succeeding activities.

The calculation will permit an activity to be split into a number of parts (each part being separated by periods during which the activity is not worked on) if limitations of resource availability require it. The following exceptions can however be specified.

i. *Non-split*. Must be scheduled continuously.
ii. *Minimum Split* (+ time T). Can be split but segments must not be less than T.
iii. *Consecutive Start* (+ time T). The first T periods of this activity must be scheduled immediately following the previous activity.
iv. *Force Finish* (+ time T). The last T periods of this activity not to be split but to be worked continuously.
v. *Consecutive Finish* (+ time T). The last T periods of this activity not to be split and the succeeding activity to commence immediately this activity is finished.

Taken in combination with the methods of specifying activity resource requirements, these alternatives provide a realistic way of specifying the manner in which groups of activities are to be performed.

The amount of each resource available at each time period can be set at two different levels: *normal* level representing the preferred maximum rate of resource usage and *threshold* level representing a higher level which can be approached but not exceeded. The threshold level will usually represent overtime working, sub-contracting, or other secondary forms of increasing capacity.

For convenience of input data preparation, various shorthand methods are used to specify the multiple combination of level and time for each resource. In addition to the normal method of stating level and duration, other notations include resource cycles, suspended cycles (e. g. interrupted by separately specified holiday periods) and combinations of cyclic and variable resources.

Resource Availability—Pool Resources

Normal and threshold resources are levels set at each particular period of scheduling time. *Pool* resources, on the other hand, represent resource levels which are not conditioned by time.

Another definition of pool resources is that they are those resources which can be carried forward from one scheduling period to the next. For example, manpower is only available for the time period specified. If 10 men are allocated to a project for two periods, the fact that only six are used in the first period does not mean that 14 will be available for the second period, If, on the other hand, the resource represented is bags of cement then 14 would be available in the second period because the unused quantity can be carried forward.

This method of defining resource availability can be used with advantage in connection with all resources which are transferable through time. It is, therefore, particularly suitable for scheduling space, materials, and money.

For example, where specific sums of money are allocated to a project, rates of expenditure on individual activities may be scheduled against a finance pool which is progressively diminished and scheduling halted when funds are exhausted. If progress payments are envisaged, then the pool will be replenished at intervals corresponding with the date of payment.

Another important application of pool resources concerns work scheduling where space is restricted, i.e., in maintenance workshops or factory assembly areas, etc. In this case, site capacity (e.g., floor area) is set as the pool limit and activity requirements scheduled against this limit.

The use of pool resource limits for consumable materials, such as cement and bricks, has been mentioned but the technique is equally applicable to nonconsumable resources, such as scaffold poles, shuttering, etc.

Progressive Feed (or Ladder) Activities

The use of progressive feed or ladder activities has become widely accepted as a simple and convenient way of depicting inter-dependent parallel activities on a planning network.

However, the method assumes a relationship between activities in the ladder chain which is not depicted by the network alone and it is necessary to introduce further data to ensure correct resource allocation. The situation is depicted in Fig. 7-3.

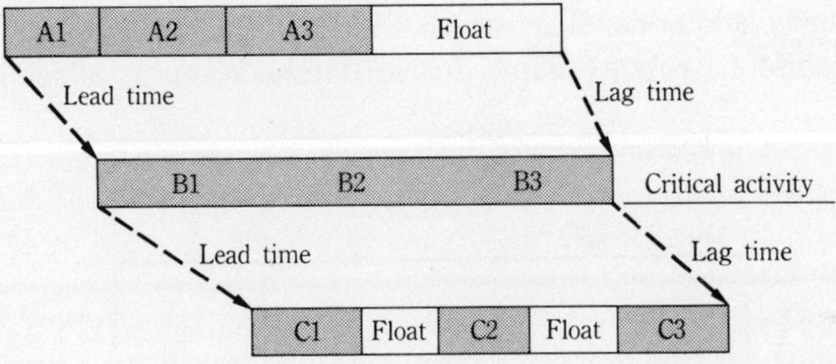

FIG. 7-3. RESOURCE SCHEDULING WITH PROGRESSIVE FEED

Here three activities A, B, and C, are interdependent and can take place in parallel subject to predetermined delays (lead and lag time).

The middle activity B is critical and has no float while the other two activities A and C have positive float. The convention assumes that work is progressively fed from one activity in the chain to the next. Therefore, if the activities are considered as discrete segments 1, 2 and 3, it is apparent that the float cannot be deployed without consideration of schedule allocated to the identical segment in the other activities in the set.

The additional information required for each activity in the chain is the preceding event of the feeding activity and the lead and lag times. The calculation then ensures proper deployment of the float (as in Fig.

7-3) and also ensures that if any activity in the chain is delayed then the appropriate delay is passed on to the interdependent parallel activities.

Scheduling—Project Duration Threshold

The use of alternative (threshold) resource levels has already been described and this technique illustrates a method of supplying the calculation with variable data to enable a greater flexibility to be embodied in the scheduling operation. Scheduling alternatives are possible, however, within a framework of project time as well; resource level and alternative project durations (threshold times) are included to further enhance this flexibility.

Two levels of increased project time are used, Project Time Change and Project Threshold Time.

The Project Time Change is the permissible increase in time (if any) beyond the project completion date calculated in the PERT time analysis. It is virtually additional float which has been allocated to the project and is therefore freely available during normal resource allocation.

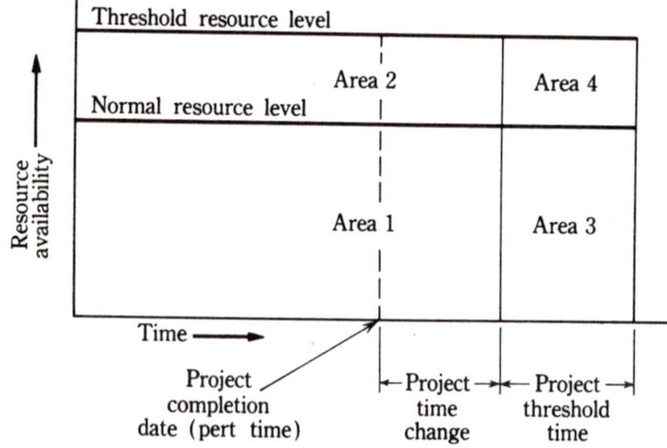

FIG. 7-4. PROJECT TIME CHANGE, PROJECT THRESHOLD TIME, NORMAL RESOURCES, AND THRESHOLD RESOURCES

The Project Threshold Time is the additional time, beyond the project time change, which may be used in dire necessity. The relationship between project time change, project threshold time, normal resources, and threshold resources is shown in FIG. 7-4.

This figure shows the alternative areas in which activity scheduling can take place and also shows the boundaries which may be set. In the case of resource availability levels they may, of course, vary with time as described earlier.

Normal scheduling would take place in Area 1 providing the amount of resource required does not exceed the normal resource level or the inability to schedule (because of non-availability of resources) does not delay project completion beyond the boundary set by the project time change.

Where either resource requirement or time requirement exceeds the agreed boundaries, then alternative schedules are possible according to whether the threshold resource is used or the threshold time is used. Thus a schedule may be calculated which uses either areas 1 and 2, or areas 1 and 3. Further non-availability of resources may cause areas 1, 2, 3 and 4 to be used. The sequence in which the boundaries of area 1 are exceeded in the case of scheduling difficulty is decided by the type of analysis which is requested. For example, a time-limited run would use threshold resources before threshold time and a normal allocation would use project threshold time before threshold resources.

In a time-limited analysis the allocation procedure is continued even if threshold resources are exceeded but the excess resource requirement is added to the alternative position which gives the least increase above the threshold resource, thus ensuring that, although the maximum permitted level is exceeded, the minimum (and hence smoothest) increase is calculated.

Where the analysis is strictly resource limited (by setting priorities) then scheduling proceeds beyond the project time threshold. In this case a "long stop" is set (to avoid the possibility of scheduling to infinity) and the calculation abandoned if this is reached.

Scheduling—Methods

Broadly, the approach to scheduling used is the conventional network method whereby a set of activities available for scheduling (i.e. all preceding activities successfully scheduled) are maintained and then these are scheduled, time period by time period, starting at the earliest date. The activities available for scheduling are arranged in a priority sequenced queue, the order of which is preset, and they are subsequently scheduled in a manner determined by a decision table (described below).

According to the resources available and the information obtained from this decision table an activity is either scheduled or "delayed." In this context "delaying" means either:

a. Delaying (for one period) an activity not already partially scheduled.
b. Splitting the activity if it is already partially scheduled in the previous time period and it is *not* designated non-split (or the minimum split time is not exceeded).
c. Restarting the scheduling of the entire activity if it is already partially scheduled and if it *is* designated non-split (or the minimum split time is exceeded).
   In this case the portion of the activity already scheduled is cancelled, and the resources added back.

For example, if when scheduling an eight-period activity a resource shortage was encountered in the seventh period, it would be either:

a. Split and scheduling ÷ 3, i.e., to leave three complete periods available for scheduling uninterrupted.

Scheduling—Decision Tables

The decision whether to schedule or to delay is made in respect of each activity by calculating four factors concerning the activity and using these four factors to reference a position in a four-dimensional maxtrix (the decision table). The four factors considered are:

1. Time situation in respect of project completion date (three cases)
   1.1 Positive total float
   1.2 Positive total float using project threshold time
   1.3 Negative total float (any condition)
2. Resource situation in respect of the requirement for the activity (three cases)
   2.1 Enough
   2.2 Enough with threshold resources
   2.3 Not enough
3. The amount by which an already partially scheduled activity would have to be unscheduled if not scheduled this period. This will be zero except for non-split activity where the amount is categorized into four ranges. The ranges can be varied but standard ranges used are:

3.1 0 to 10 periods
3.2 11 to 30 periods
3.3 31 to 60 periods
3.4 61 to infinity periods

4. Time situation in respect of project complete date if the portion of the activity already scheduled was to be cancelled and the activity rescheduled (three cases as in 1.1, 1.2, 1.3 above).

Decision table for time limited analysis – hard schedule		3. Amount Unscheduled									
		0.10 Periods			11.30 Periods			31.60 Periods			
Factor 1 Time	Factor 2 Resources	4.1. Positive total float	4.2. Positive total float at time threshold	4.3. Negative total float	4.1. Positive total float	4.2. Positive total float at time threshold	4.3. Negative total float	4.1. Positive total float	4.2. Positive total float at time threshold	4.3. Negative total float	4.1. Positive
1.1. Positive total float	2.1. Enough	S	S	S	S	S	S	S	S	S	
	2.2. Enough with threshold	D	D	S	D	S	S	D	S	S	
	2.3. Not enough	D	D	S	D	D	S	D	D	S	
1.2. Positive total float at threshold time	2.1. Enough	S	S	S	S	S	S	S	S	S	
	2.2. Enough with threshold	D	S	S	D	S	S	S	S	S	
	2.3. Not enough	D	D	S	D	D	S	D	D	S	
1.3. Negative total float	2.1. Enough	S	S	S	S	S	S	S	S	S	
	2.2. Enough with threshold	S	S	S	S	S	S	S	S	S	
	2.3. Not enou[gh]	S	S	S							

FIG. 7-5. SEGMENT OF A TYPICAL DECISION TABLE

The decision table is set up in advance for the particular type of schedule required. Using the above four factors the appropriate decision is located and extracted from the table and the scheduling (or delay) then takes place.

A segment of a typical decision table is shown in Figure 7-5.

Segment of a Typical Decision Table

In this figure it is seen that the various combinations of factors pinpoint the decision either S-Schedule or D-Delay. Shaded areas represent cases which should not occur.

In the computer program being described, nine decision tables are included. These have been devised to give the most satisfactory schedules in the following cases:

1. Aggregation at PERT time earliest date
2. Aggregation at PERT time latest date
3. Allocation, no threshold resources
4. Allocation, hard use of threshold resources
5. Allocation, easy use of threshold resources
6. An experiment table
7. Time Limited, easy use of project threshold time
8. Time Limited, hard use of project threshold time
9. Time Limited, no project threshold time

The significance of most of these will be self-evident except, perhaps, for the distinction between "hard use" and "easy use" of both threshold resources and project threshold time.

In resource allocation, project completion may be delayed if resources are not available at either the normal or threshold level. The decision to use threshold resources is therefore a function of the degree of importance attached to project completion. The "easy use of threshold resources" table will produce an earlier completion but makes greater use of threshold resources (implying that these are less important than completion date). In this case, threshold resources may be used even when total float is present. The "hard use" of threshold resources will delay the use of threshold resources except under difficult scheduling conditions (as defined by the four-dimensional decision table).

Hard and easy use of project threshold time are used in conjunction with time-limited analyses. They are similar in effect to that described for resource allocation but, of course, apply to the use of project threshold time. Easy use implies that the threshold time is entered when the slightest difficulty with scheduling is encountered and hard use means that only severe difficulties cause the time threshold to be used.

The selection of the standard tables that are shown and the distribution of decisions within the tables represent a choice of the most commonly met conditions of scheduling. However, the method is extremely flexible and it is simple to add new tables or alter standard tables as experience or special requirements dictate.

Conclusion

The developments described represent a few more steps along the path toward realistic work scheduling. The greater flexibility in specifying activity resource usage and project resource availability greatly extends the scope of network scheduling techniques. In particular, the technique of pool resources gives a new dimension to work scheduling and enables factors to be embraced which were previously excluded from this type of calculation.

The concept of project time thresholds, resource thresholds, and work continuity rules when taken in conjunction with the other features described above give an extremely comprehensive scheduling model. The refinement of correctly handling resources in conjunction with network progressive feed (ladder) activities is a useful contribution toward meeting the needs of the practitioner who desires simple methods of network expression.

The use of variable decision tables to steer the scheduling calculation gives the user of the computer a wide range of alternatives from which to choose. Taken in combination, these features provide almost infinite permutation of solution. The user is no longer in the position of having to accept the results of a fixed logic but can manipulate the model in the light of his own experience.

This complex and comprehensive computer program is also perhaps a step toward the perfect man-machine combination where man's skill and the machine's power can be harnessed together in such a way that the computer truly becomes an extension of man's intellectual powers.

This article, by H. S. Woodgate, is reprinted by permission from *Datamation*, January 1968, published and copyrighted 1968 by F. D. Thompson Publications, Inc., 35 Mason Street, Greenwich, Conn. 06830.

8

Decision Tables for Regulations

The Air Force, Army, Marine Corps, Coast Guard, and Navy are now issuing regulations in decision table format. Many civilian agencies within the government have begun to follow suit.

Why this departure from established norms? Several reasons have been offered, but the overriding one seems to be the failure of written narrative to communicate effectively.

It has been found that the decision table format forces better organization of material. The conditions and their associated actions are clearly defined in this format. And with decision tables being made up of rules to describe the various situations, one need read only applicable portions of the table to find what he is seeking. This often results in reading time being reduced by as much as 80%! Lastly, the "fog index" inherent in most narrative-type regulations is difficult to build into a decision table where one is forced to break the logic into its simplest components.

Obviously, not every regulation or directive can be reorganized into effective decision logic tables, because some directives involve no decision-making. However, a person who has become accustomed to this process of analysis will almost inevitably produce a directive or regulation of greater clarity and accuracy.

Realizing that regulations are usually based on statutes, policy statements, or similar documents in narrative form, a guide for conversion of such narrative to decision table form was published by the Air Force in September, 1965. Their technique for converting a narrative directive to decision logic tables, which originally appeared as Chapter 3 of Air Force Pamphlet 5-1-1, follows:

Restructuring a Narrative Publication to Decision Logic Tables (extract from AFM 36-10)

Approach To Conversion

Restructuring a publication from narrative to DLT format is accomplished through analysis and documentation. Analysts, writers, and subject matter specialists may develop variations of this step-by-step approach as the need, type, and size of procedure may suggest. Process steps are:

a. Review the objective.
b. Review the publication.
c. Determine major events.
d. Determine components of major events or subevents.
c. List conditions and actions (prepare grid chart).
f. Analyze grid chart.
g. Structure preliminary DLT(s)
h. Analyze drafts.
i. Prepare final DLT(s).
j. Test and evaluate.

Step-by-Step Analysis

To demonstrate, we are using a portion of AFM 36-10 which deals with officer effectiveness reporting. *Keep in mind that the subject matter of the problem text may have changed due to policy and procedure adjustment.*

A. STEP 1—REVIEW THE OBJECTIVE

This provides the drafter with a point of reference in the overall procedure. The drafter should consult with the policymakers to determine if the stated objective(s) is still appropriate. The objective and goals of the entire directive must be thoroughly understood and constantly kept in mind. In the case of AFM 36-10 the objective is simply stated as the document which establishes Air Force policy, provides instructions on the preparation, submission, review, and control of Officer Effectiveness Reports and Officer Training Reports on officers in the grades of colonel and below.

B. STEP 2—REVIEW THE PUBLICATION

Determine what is policy, what constitutes administrative-legal considerations, and what are definable procedures.

(1) Study the procedures to thoroughly understand the detail involved, regardless of the level at which the steps of the procedure are to be performed.

(2) Separate the remaining portions into policy, definitions, administrative matters, and legal references. These usually do not lend themselves to tabular presentation; however, when policy depends on alternate conditions this portion should be analyzed and worked up in the same manner as procedures.

C. STEP 3—DETERMINE MAJOR EVENTS

Most publications take a generalized approach toward describing the content. In DLTs major events must be identified, listed, given appropriate and understandable titles, and then arranged in a logical sequence. Each major event must have an objective which may or may not be indicated as such in the final product; but the objective should be kept in mind and completely understood. Of the several major events in AFM 36-10 (some of which are: who will prepare the report, under what conditions will the report be referred to the officer reported on, indorsement of reports, and distribution of completed reports), that procedure of indorsement of reports is used to illustrate what happens to a major event under analysis. This narrative is reproduced in Fig. 8-1.

D. STEP 4—DETERMINE COMPONENTS OF MAJOR EVENTS OR SUBEVENTS

Consider further dissecting each major event into component parts or subevents. The procedure here is similar to that for identifying and listing major events in Step 3.

FIG. 8-1. EXTRACT FROM AFM 36-10.

5-3. Who Will Indorse the Effectiveness Report.

This responsibility is confined to the chain of command and will not be delegated. The report will be indorsed initially by the person immediately supervising the reporting official.

a. If the reporting official is junior in rank or grade to the officer being reported on, the indorsing official will cite the special order or other document effecting the duty assignment of the reporting official.

b. If an indorsing official is junior in rank or grade to the reporting official or a prior indorsing official, he will cite the special order or other document effecting his duty assignment.

c. When both the reporting and indorsing officials are junior in rank or grade to the officer reported on, the report will be additionally indorsed by the immediate supervisor of the indorsing official.

d. Outstanding or referral reports (as defined in Chapter 4) on officers below the grade of colonel will be additionally indorsed by one of the following:

(1) A USAF general officer.

(2) The appropriate general *or flag officer* in joint activities.

(3) A colonel occupying a manpower authorization document position of general officer or serving as commander of a wing, or as a commander at equivalent or higher echelon.

(4) Civilians in grade GS-16 and above, whose duties have been designated as being equivalent to those performed by a general officer.

e. Outstanding or referral reports (as defined in Chapter 4) on colonels will be additionally indorsed by one of the following:

(1) A USAF general officer.

(2) The appropriate general of flag rank in joint activities.

(3) Civilians in grade GS-16 and above, whose duties have been designated as being equivalent to those performed by a general officer.

Note: When the normal indorsing official falls into one of the categories cited in paragraphs d and e above, the additional indorsement is, although permitted, not required.

f. The additional indorsing official required in d or e above will review the ratings and comments of the reporting and indorsing officials for completeness and impartiality and indicate his agreement or disagreement with the report. Even though the additional indorsing official may not have personal knowledge of the officer reported on, he can accomplish an effective review of the report to determine its qualitative adequacy. This review serves both the purposes of quality control over individual reports and of control over rater and indorser tendencies to overrate. The mandatory additional indorser should unhesitatingly reject poorly prepared reports and downgrade ratings that are not substantiated or reflect unacceptable inflationary practices. If he does not agree with one or more ratings he may place his initials (without encircling them) in the appropriate boxes. Any disagreement with ratings or comments will be substantiated by specific comment. In addition, include

FIG. 8-2. INITIAL GRID CHART

	CONDITIONS	ACTIONS A Indorsed initially by immed super-visor of rept official	B Indorsing official will cite other SO or document eff duty asgmt of rept ofl	C Indorsing ofl will cite SO or other document eff his duty as-signment
1	If reporting official is junior to officer reported on		X	
2	Indorsing ofl is junior to rept offl or prior ind official			X
3	Both rept & ind ofl are jr to off reported on			
4	If rept is outstanding or refrl on officers below grade of colonel			
5	If rpt is outstanding or referral on a colonel			
6	If outstanding or refrl and normal ind ofl is either note 1 or 2			
7	If outstanding or refrl and addn ind ofl is either note 1 or 2 and agrees			
8	If outstanding or refrl and addn ind ofl is either note 1 or 2 and disagrees			
9	Any ofl superior to indorsing ofl in chain of command			

any information that will contribute to a more complete report.

g. Any official who is superior to the indorsing official in the chain of command may attach an additional indorsement (AF Form 77a or 707a, as appropriate) to any report, provided it adds substantive information about the officer's performance; however, he will not place his initials in any section to indicate disagreement with the contents of the report. Disagreement will be noted in the body of the indorsement, with any comments which will add to the objectivity of the report.

Note: Paragraphs f and g outline general criteria and because of this will be carried in the title sheet when this regulation is fully reconstructed using DLTs.

D	E	F	G	H	I	J
Addn indorse by immed super-visor of indorsing official	Addn indorsed by gen off (see note 1)	Addn indorsed by gen off (see note 2)	Initial boxes & substan-tiate by specific comment	May add indorse-ment if it adds substan-tive info	Addn in-dorse-ment permitted	Forward complete report
X						
	X					
		X				
					X	
						X
			X			
				X		

E. STEP 5—LIST CONDITIONS AND ACTIONS (PREPARE GRID CHART)

List the conditions and actions in a random but complete manner for each major event or subevent. There is no particular method for listing conditions and related actions; however, using a grid chart (Fig. 8-2) clearly and effectively displays these relationships. Graph paper (22 x 17″) or similar types of ruled paper are useful in preparing grid charts.

(1) In listing the conditions and actions on the grid chart there may be duplication. Do not try to eliminate this duplication at this point unless it is obvious that the condition or action is identical to one previously stated but worded differently. Do not read into the portion of the publication being worked on procedures which are implied but not stated; these implications or omissions will be pointed out later in the analysis. Items which are not clear may be indicated on the grid chart itself or listed separately for later clarification.

(2) When "gridding," references to other sections in the publication or other publications may be included in the narrative. Note these references and keep them with the grid chart (on 3 x 5″ cards, for example) with a statement pointing out the reference and relationship. Then check the matter to which the references are cross-referenced. Including the cross-referenced data in the grid chart may do away with the need for cross-referencing.

(3) Figure 8-2 lists the condition and action relationships extracted verbatim from the narrative in Fig. 8-1. (Some obviously clear abbreviations were used.) The introductory sentence of the text to be translated reads in part, "The report will be indorsed initially by the person immediately supervising the reporting official." It is identified as an action and is listed in the action entry part of the chart (across the top of column A). Random listing of conditions and actions proceeds. Reference the entry, "*If* reporting official is junior to the officer reported on." This is identified and listed as a condition on line 1. The related action to this condition is the remainder of the next sentence, and is entered as an action in column B. Line 1 and column B are connected by an "X" at their intersection to show the relationship between this condition and action. This process is continued until all of the narrative has been transferred to the grid chart. As the narrative is being transferred to the grid chart it should be checked off to insure a complete transfer.

(4) Certain data may be exceptions, or, in some cases, may pertain to the whole of the chart or several conditions or actions. Such data

may be treated as notes, and thus can become part of the table instead of being lost or missed in the narrative which is left after all tabular work has been completed. Reference to the notes must be placed on the grid chart (columns E and F of Fig. 8-2); the notes themselves (Fig. 8-3) must be placed on or attached to the grid chart.

FIG. 8-3. SAMPLE NOTES TO DLT

NOTE 1: OUTSTANDING or REFERRAL reports (defined in Chapter 4) on Lt. Colonels and below will be additionally indorsed by one of the following:

(1) A USAF general officer.

(2) The appropriate general of flag rank in joint activities.

(3) Civilians in grades GS-16 and above, whose duties have been designated as being equivalent to those performed by a general officer.

(4) A colonel occupying a manpower authorized document position of general officer or serving as commander of a wing, or as a commander at equivalent of higher echelon.

NOTE 2: OUTSTANDING or REFERRAL reports on colonels will be additionally indorsed by one of the following as listed in (1), (2), or (4) of note 1.

F. STEP 6—ANALYZE GRID CHART

After transcribing all narrative to the grid chart, the chart must be analyzed. This is a crucial point.

(1) Check the grid chart against the narrative to insure that all situations, actions, notes, and internal and external cross-references have been transcribed, thus eliminating further reference to the narrative.

(2) By using the grid chart it is possible to relate each condition to each action and investigate the applicability of the condition to the action and vice versa. The analyst may use either a symbol (such as the "A" used in Fig. 8-4) or a different color pencil in his analysis notation so that he can distinguish the analytic portion from what was actually extracted from the basic document. Figure 8-4 shows that by analysis the action in column A is common to all conditions. This points out that possibly this column can be combined later to reduce the table size and wordage. Next, each condition is compared to the action in column B. This process is repeated for each condition and action to find common relationships by asking the question, "If a condition exists, must, could, or should the action be taken?"

(3) A condition or set of conditions must be traceable to its related action(s) and an action must have its related condition(s). When there

is no condition for a specific action or no action for a specific condition, the grid chart is not complete.

(4) When only a portion of a condition exists another action would be applicable, or, conversely, only a portion of an action would be applicable for a stated condition. This occurs when the condition or action has not been divided far enough. Proper separation should be entered on the grid chart.

(5) Another point to look for is action statements which contain

FIG. 8-4. ANALYZED GRID CHART

ACTIONS

	CONDITIONS	A	B	C	D
		Indorsed initially by immed super-visor of rept official	Indorsing official will cite SO or other document eff duty asgmt of rept ofl	Indorsing off will cite SO or other document eff his duty as-signment	Addn indorsed by immed super-visor of indorsing official
1	If reporting official is junior to officer reported on	A	X		
2	If indorsing ofl is junior to rept ofl or prior ind official	A		X	
3	Both rept & ind ofl are jr to off reported on	A	A	A	X
4	If rept is outstanding or refrl on officers below grade of colonel	A	A	A	A
5	If rept is outstanding or referral on a colonel	A	A	A	A
6	If outstanding or refrl and nor-mal ind ofl is either note 1 or 2	A			
7	If outstanding or refrl and addn ind ofl is either note 1 or 2 and agrees	A			
8	If outstanding or refrl and addn ind ofl is either note 1 or 2 and disagrees	A	A	A	
9	Any ofl superior to ind ofl in chain of command	A	A	A	A

words such as *if* or *when;* these are conditions and should not be indicated as actions.

(6) When more than one condition results in an action or series of actions, a multiple relationship exists. By associating the Xs and As on the grid chart, combinations of certain conditions lead to combinations of related actions. These multiple relationships point out the possibility of consolidating a number of conditions into a single condition, or a number of actions into a single action.

	F	G	H	I	J
ddn dorsed gen (see te 1)	Addn indorsed by gen off (see note 2)	Initial boxes & substantiate by specific comment	May add indorsement if substantive info	Addn indorsement permitted	Forward complete report
X					
	X				
				X	
					X
		X			
A	A	A	X		

TABLE 8-1. INDORSEMENT OF ALL OERS

	A	B	C	D	E
RULE	If type of report is	and reporting official is	and indorsing official is	and additional indorsing official is	and the grade is
1		jr to off being rated			
2			jr to rept ofl or prior ind ofl		
3		jr to off being rated	jr to off rept on		
4	outstanding or referral				lt col & below
5	outstanding or referral				col
6	outstanding or referral		USAF gen off (note 1 or 2)		
7	outstanding or referral			USAF gen off (note 1 or 2) agrees	
8	outstanding or referral			USAF gen off (note 1 or 2) disagrees	
9	any			any superior in chain command	

G. STEP 7—STRUCTURE PRELIMINARY DLT(s).

Since the data have been transcribed and analyzed, the next requirement is to construct a DLT. The first consideration is to develop meaningful and appropriate titles for the condition and action stubs. All condition and action entries must now be transferred from the grid chart to the DLT, which has the quadrant format.

F	G	H	I	J
initially indorsed by immed supervisor	ind ofl will cite SO or other doc eff	additional indorsement by	initial blocks	other
reporting official	duty asgmt of rept ofl			
reporting official	his duty assignment			
reporting official	duty asgmt of rept ofl & his duty asgmt	immed supv of ind ofl		
reporting official		USAF gen off (note 1)		
reporting official		USAF gen off (note 2)		
reporting official		permitted		
reporting official				forward completed report
reporting official			per-mitted	must be sub-stantiated by comment
reporting official		permitted	no	comments must add info on off being rated

(1) *Determine Condition and Action Stubs*

The stub entries are determined by generalizing the conditions and actions listed on the grid chart, where practical. (Refer to Table 8-1.) Analysis revealed more than one type of report, a common condition relationship under the heading of "reporting official," more than one condition referred to "indorsing official," and there was also a "grade" consideration. As regards the action stubs, more than one reference was made to "initially indorsed by immediate supervisor," "indorsing official will cite special orders or other document effecting," and "additionally indorsed," etc. In this table the stub titles appear to be meaningful.

(2) *Apply Condition and Action Entries to Table*

Once the condition and action stub titles have been determined and entered on the DLT, the remainder of the condition and action statements will be transcribed from the grid chart. For example, the first condition reads, "If reporting official is junior to officer reported on." This, as the first condition, fits under the condition stub entry "reporting official." It is placed in rule 1 and identifies itself with condition 1 on the grid chart. This numerical integrity should be maintained because it will cross-reference to the grid chart and assist in checking for completeness of the DLT and otherwise assist in further analysis. Next, transfer the remaining portion of the action statements related to the condition from the grid chart to the action entry of the preliminary DLT. Follow this procedure for each condition and action until all conditions and actions are transferred.

H. STEP 8—ANALYZE DRAFTS

Often this is the most revealing step in the process. The narrative has been transcribed to the grid chart and the data on the grid chart have been arranged on the first DLT. Nothing has been deleted from the original text. The only changes are the format in which the information is displayed and the possible addition of explicit interpretations of data which were previously implied in the narrative. It could very easily be returned to the narrative at this point and result in a more complete narrative in most cases, except that there has not been a check made for completeness or proper sequence. In most cases the data contained in the first DLT are not complete nor are they in the proper sequence.

(1) Analyze the condition portion of the table to determine if a primary condition exists upon which all other conditions are based. This is a single condition stub, with its various entries, upon which all other conditions depend. In Table 8-1, column A is the primary condition column, since all officer effectiveness reports will meet one of these conditions. The blanks in rules 1, 2, and 3 indicate that it can be any type of report. The remaining conditions should be indicated in the most logical sequence. If there are two or more primary conditions

which are exclusive of each other, it indicates that more than one table will be required. For example, refer to Table 8-1 again; ignore column A and look at columns B, C, and D. In rule 3, column C, the entry depends on the entry in rule 3, column B: therefore, these two condition columns depend on each other in relation to the actions which must be taken. Reference Column D, the only condition entries listed, are in rules 7, 8, and 9. They are independent of both columns B and C; therefore, this indicates that two tables would be required if these three columns were the only condition columns in the table.

(2) Examine the action portion of the original draft DLT from the standpoint of the proper sequence in which actions should be taken. The sequence of actions insures continuity and logical progress through the procedures, and must be considered carefully throughout the development phase. When based on certain conditions, an action or actions will be taken. Before final action can be taken, however, still other conditions must be considered after the first action is taken. This determines the requirement for more than one table. For example, let's look at Table 8-2. After conditions A, B, and C are satisfied in rules 1

TABLE 8-2. INDORSEMENT OF ALL OERs

RULE	A	B	C	D
	Reporting official	indorsing official	OER is outstanding or referral	additional indorsed by (gen off) (note 1)
1	senior to off rated	senior to rept ofl	no	
2	senior to off rated	senior to rept ofl	yes	who agrees
3	senior to off rated	senior to rept ofl	yes	who disagrees

and 2, condition D, which reads "additionally indorsed by a general officer," cannot be reached at this point because his agreement or disagreement could not be known. The drafter is made aware that all conditions under structure could not be satisfied, thus the proper action could not be taken. The need for more than one table to flow the logic and the procedure is quite evident. *There is a tendency to cram as much data as possible into one table. This should be avoided since it tends to confuse the user and complicate the procedure.*

(3) All tables must be carefully analyzed to insure that all the logic

associated with the process being documented has been provided. This includes a complete listing of all conditions and actions, all multiple relationships and all alternatives. For example, refer to Table 8-1 again. In column B the only mention of the reporting official is to those who are "junior to the officer being rated." This is an exception since approximately 99% of all reporting officials will be senior to the officer being rated. This has been omitted from the original directive, but it

TABLE 8-3. INDORSEMENT OF ALL OERS

RULE	A	B	C	D
	If reporting official is	and indorsing official is	and OER is outstanding or referral	and the grad of off reporte on is
1	senior to off reported on	senior to rept ofl	no	
2	senior to off reported on	senior to rept ofl	yes	colonel
3	senior to off reported on	senior to rept ofl	yes	lt col & belo
4	senior to off reported on	junior to rept ofl	no	
5	senior to off reported on	junior to rept ofl	yes	colonel
6	senior to off reported on	junior to rept ofl	yes	lt col & belo
7	junior to off reported on	senior to rept ofl & off rept on	no	
8	junior to off reported on	senior to rept ofl & off rept on	yes	colonel
9	junior to off reported on	senior to rept ofl & off rept on	yes	lt col & belo
10	junior to off reported on	junior to off reported on	no	
11	junior to off reported on	junior to off reported on	yes	colonel
12	junior to off reported on	junior to off reported on	yes	lt col & belo

should be included. There is another relationship omitted when column B is related to column C. All reports must have an indorsing official and a reporting official, and the actions taken are based on the relationship of the reporting and indorsing officials to the officer being rated. The blanks in columns B and C should be filled in and their actions identified as in Table 8-3. Again keep in mind the questions used in

	F	G	H
-en OER will -e indorsed by -nmed supv	and the ind ofl will cite the SO or other doc eff	and additional indorsement will be by	GO TO
-porting official			Dist Table
-porting official		USAF gen off or equiv (see note in Fig. 8-3)	Table 8-6
-porting official		USAF gen off or equiv (see note 1)	Table 8-6
-porting official	duty asgmt of ind official		Dist Table
-porting official	duty asgmt of ind official	USAF gen off or equiv (see note 2)	Table 8-6
-porting official	duty asgmt of rept official	USAF gen off or equiv (see note 1)	Table 8-6
-porting official	duty asgmt of rept official		Dist Table
-porting official	duty asgmt of rept official	USAF gen off or equiv	Table 8-6
-porting official	duty asgmt of rept official	USAF gen off or equiv (see note 1)	Table 8-6
-porting official	duty asgmt of rept ofl & ind ofl	immed supvr of ind ofl	Dist Table
-porting official	duty asgmt of rept ofl & ind ofl	immed supvr of ind ofl & USAF gen off (note 2)	Table 8-6
-porting official	duty asgmt of rept ofl & ind ofl	immed supvr of ind ofl & USAF gen off (note 1)	Table 8-6

analyzing the grid chart: "If a condition exists, must, could, or should the action be taken?" Another way to look at the conditions stated is to determine if the reverse of these conditions are indicated or if they could exist; if they could exist they should also be litsed and their actions determined and indicated.

(4) In Table 8-3 we cover all the possible conditions and indicate the applicable actions required, but the table is unwieldy and almost overwhelming. Simplicity is needed for correct interpretation, clarity, and efficient utilization. This table further reflects that the number of variables involved affects the ability to clearly isolate applicable conditions and their related actions. It takes too long to read the rules and find the appropriate action. By analysis of column C, Table 8-3, it is evident that only ⅓ of the table is required to process the normal case (when the report is *not* outstanding or referral). Possibly 97% or more of the cases fall in this category. Also, column D has a direct relationship to the exception cases in column C; actions based on these conditions are for an additional indorsement of the report. By removing these two columns from this table and placing them in a separate table we are left with Table 8-4, which is small, simple to interpret, and easy to use.

(5) DLTs must reflect the ease of use, size, numbers, sequence, and the procedure interrelationships needed to obtain the objective of improved procedures. Consider the individual who must use the table. *Insure that wordage and content are at a realistic reading level.* Confine tables to 8 x 10½" printed page when possible. More than one table can be placed on a page. The number of tables in a procedure is not restricted; the goal is to simplify procedures. When more than one table is used in a procedure, and sequence of tables is not apparent, proper GO TO instructions must be used (reference column H of Table 8-3, Figure 8-5, and Table 4-12).

(6) The frequency of occurrence of a particular condition or set of conditions or actions must be considered in establishing the relative positioning of each rule on the table. If other than outstanding or referral report is the most common situation faced by the user, then his set of conditions and actions must be at the top of the table (rule 1, Table 8-3). Table 8-4 reflects this by considering that most reports are rendered by reporting officials who are *senior* to the officer rated, and indorsed by officials *senior* to reporting officials. Thus, the *junior* in grade and rank situation is relegated to a lower position in the table.

TABLE 8-4. INDORSEMENT OF ALL OERS

RULE	A If reporting official is	B if normal indorsing official is	C initial indorsement by immediate supervisor	D ind ofl will cite SO or other document effecting	E additional indorsement	F GO TO Table 8-7
1	senior to officer rated	senior to rept ofl	reporting official			X
2	senior to officer rated	junior to rept ofl	reporting official	duty asgmt of ind official		X
3	junior to officer rated	sr to off rated and rept official	reporting official	duty asgmt of rept official		X
4	junior to officer rated	junior to officer rated	reporting official	duty asgmt of rept and ind official	immed supv of ind ofl	X

I. STEP 9— PREPARE FINAL DRAFTS

Figure 8-5 is the final structuring of the sample text. These three tables and the lead-in narrative, which is policy in nature, are converted from the narrative. The tables do not represent the only refinements possible, since the policymaker has within his primary interest the authority to modify and structure better procedures. However, they are a proper solution from the standpoint of proper analysis and structure.

(1) The review of the grid chart (Figs. 8-2, 8-3, and 8-4) and Tables 8-1, 8-2, 8-3, and 8-4 point out that two general conditions dealing with junior officer and senior officer fell under the heading of reporting official and normal indorsing official. The actions centered on the initial indorsement, the citing of the special order in other than routine instances, additional indorsement, and table/procedure sequencing. In addition, the separation of the normal run of reports from those carrying referred or outstanding identification appeared practical.

(2) Table 8-6 demonstrates the use of notes. Notes should be read *only* when the problems-solving rule so directs. *Every table that has notes must have a reference within the table to the notes.* In this instance the note was rewritten by the drafter, the result being a more simple presentation mainly through the elimination of repetition.

(3) Table 8-7 deals with procedures involving additional indorsement—yes or no. The report flows logically through initial indorsement and additional indorsement actions required due to exceptional circumstances.

J. STEP 10—TEST AND EVALUATE

The drafter must now make sure the final drafts are complete and adequate, and that they reach the objective. He must reapply the analysis rules in step 8. He will then simulate "live" situations and apply them to the tables to insure clarity, efficiency, and completeness. Using both typical and unusual situations may reveal the need for modification.

Who Will Indorse the Effectiveness Report:

a. Responsibility is confined to the chain of command and will not be delegated.

b. Any official who is superior to the indorsing official in the chain of command may attach an additional indorsement, provided it adds substantive information about the officer's performance. He will not place his initials in any section to indicate disagreement with the contents of he report. Disagreement will be noted in the body of the indorsement, with any comments which will add to the objectivity of the report.

c. Comments of more than one additional indorsing official may be included on a single indorsement sheet. In this case authentication by the preparing official in the format shown in Section II of the indorsement sheet will immediately follow each indorsement.

d. Mandatory indorsing officials will be as indicated in the following tables:

Fig. 8-5. Final DLT version of fig. 8-1.

Table 8-5. Indorsement of all OERs (extended entry—horizontal format)

RULE	A	B	C	D	E	F
	If reporting official is	and the normal indorsing official is	then initial indorsement will be by immediate supervisor of	and indorsing official will cite special orders or document effecting	and be additionally indorsed by	GO TO Table
1	senior to officer rated	senior to reporting official	reporting official			8-8
2		junior to reporting official	reporting official	duty assignment of indorsing official		8-8
3	junior to officer rated	senior to officer rated and reporting official	reporting official	duty assignment of reporting official		8-8
4		junior to officer rated	reporting official	duty assignment of reporting and indorsing official	immediate supervisor of indorsing official	8-8

TABLE 8-6. ADDITIONAL INDORSEMENT

RULE	A	B	C	D
	If report is outstanding or referral	and if normal or additional indorsing official is	then additional indorsement	GO TO
1	no			Distribution Table
2	yes	below USAF general officer or equivalent (see note)	required by USAF general officer (see note)	Table 8-7
3	yes	USAF general officer or equivalent (see note)	permitted	Distribution Table

TABLE 8-7. ADDITIONAL INDORSEMENT BY A GENERAL OFFICER

RULE	A	B	C	D
	If general officer's indorsement	initial in appropriate blocks is	and indorsement must include	GO TO Distribution Table
1	agrees			X
2	disagrees	permitted	specific substantiation of disagreement	X

9

Simulation with Decision Tables

The area of simulation can be divided into three parts: machine, man-machine, and machine independent simulations. Within the field of machine simulation, the area which has lately received most attention is that of digital simulation.

INTRODUCTION

Programming a problem for digital computers has been aided considerably by the development of special purpose programs, assemblers, interpreters, and higher level languages. The basic purpose of all these "languages," and the "software packages" used to implement them (compilers, documentation, diagnostic aid, etc.), is to provide the definer of the problems as well as the programmer with a means for communicating with the machine without using machine language. The packages take care of many of the mechanical details of programming: memory location assignments, input-output commands, and relationships among program segments. The languages also help the problem definer and the programmer communicate with each other.

Computer simulation has come into increasingly widespread use in the study of the changing behavior of systems. Alternatives to the use of simulation are mathematical analysis, experimentation with either the actual system or a prototype of the actual system, or reliance upon experience and intuition. All, including simulation, have limitations. Mathematical analysis of complex systems is very often impossible; experimentation with actual or pilot systems is costly and time consuming, and the relevant variables are not always subject to control. Intuition and experience are often the only alternatives to computer simulation available, but they can be very inadequate.

Simulation problems are characterized by being mathematically intractable and having resisted solution by analytic methods. The problems usually involve many variables, parameters, functions which are not well behaved mathematically, and random variables. Thus, simulation is a technique of last resort. Yet much effort is now devoted to "computer simulation" because it is a technique that gives answers in spite of its difficulties, costs, and time required.

There are many simulation languages available for use on the computer. Some of them deal with continuous change models which can be simulated on digital computers by using finite-difference equations which, in the limit, approach the differential equations of continuous flow.

Others are discrete-change models in which the changes in the state of the system are conceptualized as discrete rather than continuous. Systems are idealized as network flow systems and are characterized by the following:

1. The system contains "components" (or "elements" or "subsystems") each of which performs definite and prescribed functions;
2. Items flow through the systems, from one component to another, requiring the performance of a function at a component before the item can move on to the next component;
3. Components have finite capacity to process the items and, therefore, items may have to wait in "waiting lines" or "queues" before reading a particular component.

The main objective in studying such systems is to examine their behavior and to determine the "capacity" of the system; e.g., how many items will pass through the systems in a given period of time as a function of the structure of the system. The analytical techniques which may be used to solve such problems are queueing theory and stochastic processes. Examples of problems which have been formulated and studied as discrete change models are job shops, communications networks, logistics systems, and traffic systems.

The computation in this type of simulation consists to a large extent of keeping track of where individual items are at any particular time, moving them from waiting line to component, timing the necessary processing or functional transformations, and removing and transporting the items to other components or waiting lines. The result of a simulation "run" is a set of statistics describing the behavior of the simulated system during the run. Special packages have been prepared for certain specific applications. To use these, the user supplies parameter values, data, and control values to adapt the program to his own model.

Since simulation of discrete-change models involves computations which are common to many models, a number of languages and packages have been designed for formulating and writing a program for any discrete-change model.

When a simulation language such as SIMSCRIPT or GPSS-II is not available to the programmer, he has to rely on his own ingenuity and programming ability if he is attempting to simulate some type of complex system. The more complex a system becomes, the harder it would be to program using a general language, such as FORTRAN. For this reason a higher level of simulation language was developed.

Decision tables can be used to simplify a situation in which complex logic plays a role. Thus, if there is a very complex simulation to program and there is no higher level language available to code the problems, the use of decision tables might simplify the coding of the simulation to a certain degree.

The objective of this chapter then is twofold: first, to present the technique of simulation with the use of decision tables, and secondly, to evaluate this technique in terms of time saved, ease of understanding, and future use. The specific application of decision tables to simulation will be in queueing systems, both simple and complex.

Queueing Systems

Queueing systems range from a simple, single service counter, single queue, to the more complex situation including many service counter sets, each containing a number of queues. The models used here are experimental in nature. They can be varied quite easily with respect to structure.

MODEL 1—SINGLE SERVER, SINGLE QUEUE

The first simulation model used is illustrated in Fig. 9-1. This is a simple queueing model with one service counter and one queue for that service counter. Arrivals are generated at a constant interval of time. For every arrival which enters the system a processing time is generated and assigned to that arrival. From this processing time we can determine the departure time of the arrival and place it either "in process" or in the queue, depending upon the status of the system.

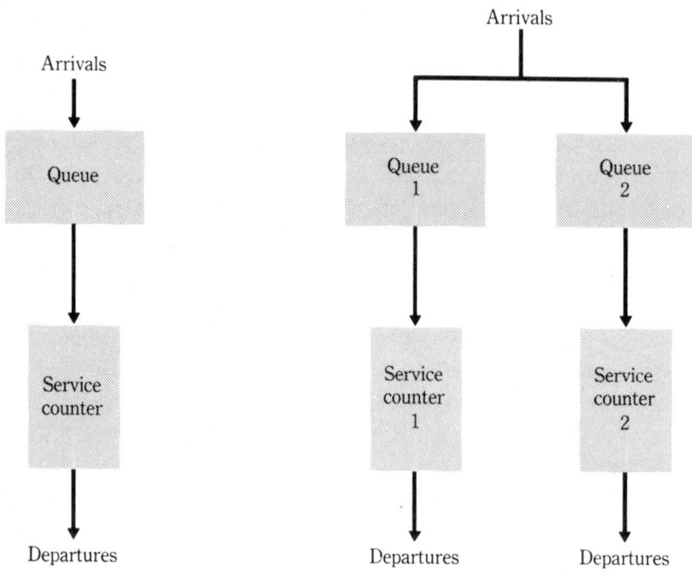

FIG. 9-1 FIG. 9-2

MODEL 2—TWO SERVERS, TWO QUEUES

The second queueing model used is more complex than the first. It is illustrated in Fig. 9-2 and has two service counters and two queues, one associated with each service counter. Arrivals in this system are also constant (every five units of time). When an item (or person) arrives, both service counters must be checked to see if an item is in process. If both of the service counters are empty then the arrival can proceed to either counter and be processed—a process time is generated for each arrival upon its arrival to determine its departure time. If only one service counter is empty the arrival can proceed to that counter and be processed. If both service counters are occupied, a decision has to be made as to which queue the arrival should be assigned. This decision is based on specific rules followed in this order:

1. Assign arrival to queue with the least items in it.
2. If both queues contain the same number of items, assign the arrival to the queue which has the earliest departure time between the last items in the queues.

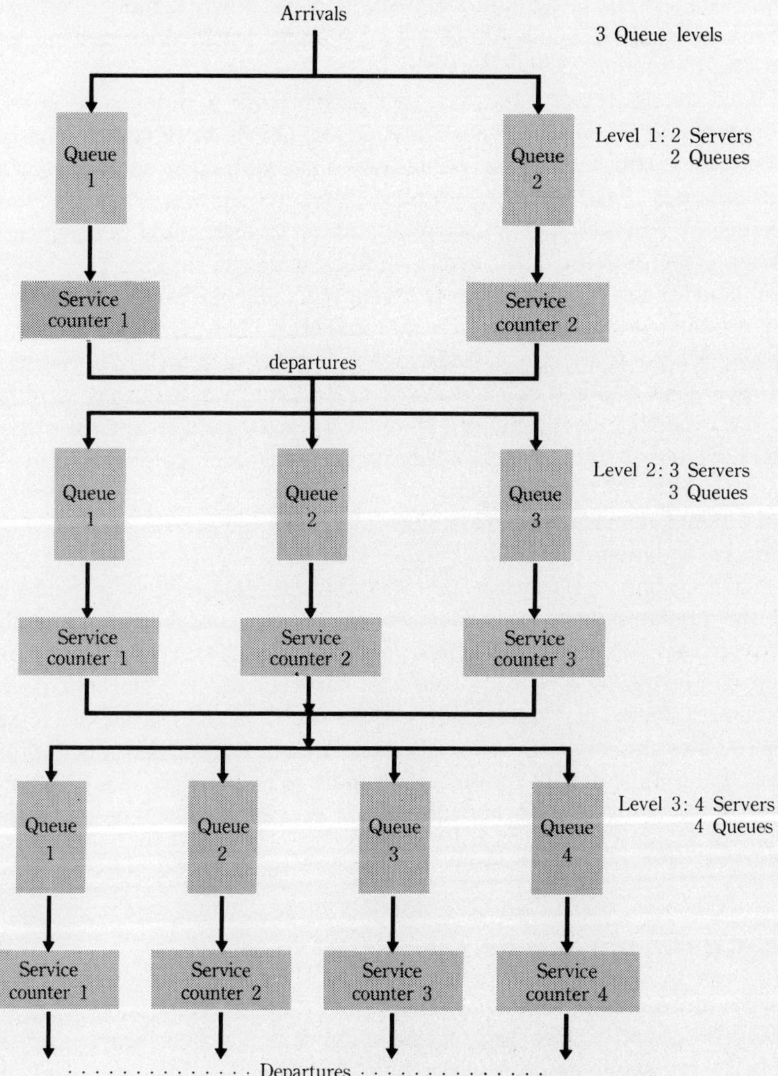

FIG. 9-3

MODEL 3—COMPLEX QUEUEING SYSTEM

The third queueing model used is a complex model structurally, as well as from the programming point of view. Such a model can very easily have many applications, e.g., job shop, truck, dock system, and so on. The model is illustrated in Fig. 9-3.

This model consists of three queue sets with a number of service counters and individual queues in each set. Queue set Number One has two service counters and a queue associated with each counter. Queue set Number Two has three service counters and three queues, and queue set Number Three has four service counters and four queues. When the processing of an item is finished at queue set One (1st level), the item passes on to the next queue set, and so on until the item leaves the system through queue set Three. This is a very flexible model, since by changing the parameters which determine the number of queues in a queue set, the whole model can be rearranged resulting in the capability to run many different tests on the same general model. At each queue level certain priority decision rules could be inserted to determine the path or behavior of a particular item in the system. No attempt is made in these models to govern the arrival or processing time by equations.

Each queue level, except the first, is dependent upon the behavior of the previous queue set. The first queue set is dependent upon the arrivals that come at a constant interval of time. The second queue set depends on the number of departures from queue set One, and hence queue set Three depends on the number of departures from queue set Two. Once this model has been set up, it does not require much effort to expand this model to one with more queue levels. This can be accomplished by duplicating the actions (decision table) of queue set Two as many times as desired.

Description of Decision Tables Used to Simulate

The decision rules which govern the behavior of a queue can be easily described by the use of a decision table. A flow chart is probably easier to understand if the number of queues in a queue set is relatively small. (At this point we are talking about the computer program and the flow chart associated with that program which describes

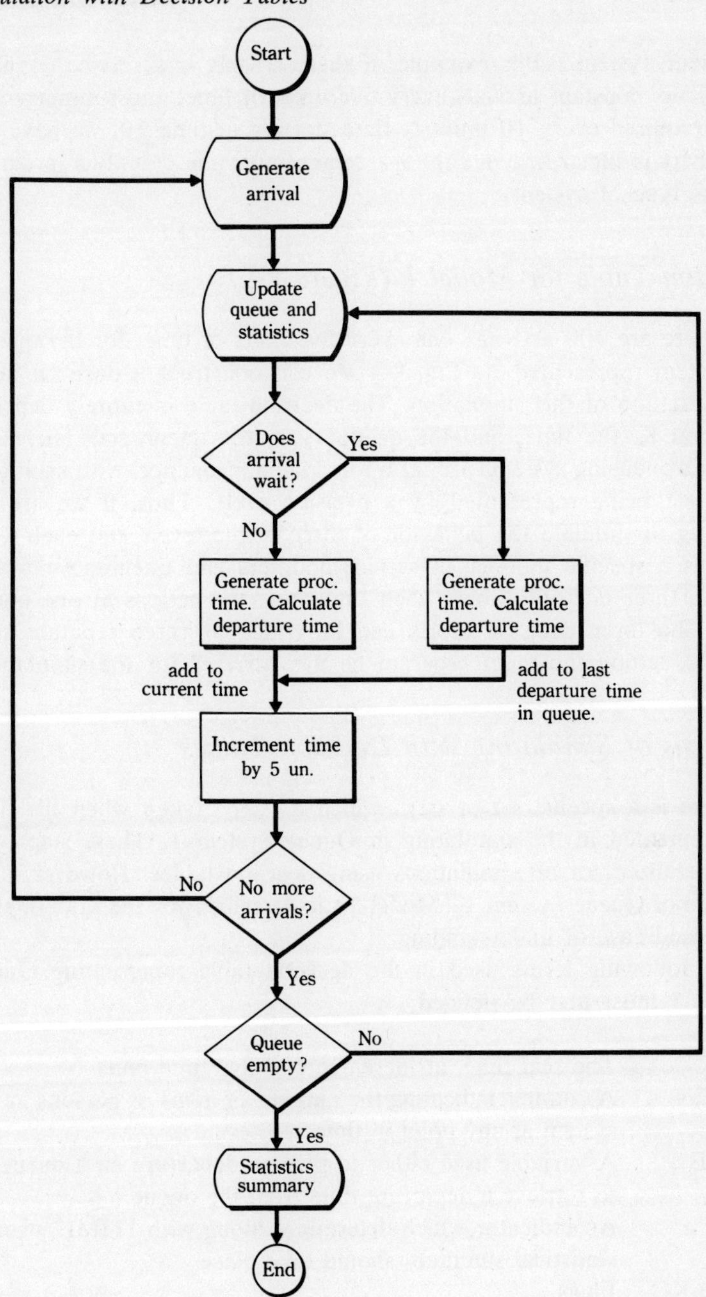
Start
Generate arrival
Update queue and statistics
Does arrival wait?
Yes
No
Generate proc. time. Calculate departure time
Generate proc. time. Calculate departure time
add to current time
add to last departure time in queue.
Increment time by 5 un.
No
No more arrivals?
Yes
Queue empty?
No
Yes
Statistics summary
End

FIG. 9-4

the queue system.) For example, if there is only one service counter, one queue, constant arrivals every five units of time, and summary statistics required every 10 units of time starting at time 10, we have the flow chart is illustrated in Fig. 9-4 representing the computer program for this type of system.

Decision Table for Model 1 (Figure 9-5)

If there are 100 arrivals, one every five units of time, for the queueing system represented by Fig. 9-4 we can construct a decision table representation of this simulation. The decision table is entirely dependent upon K, the time, and the queue(s) in the queue set. The more complex queueing systems are set up in a similar manner, with each new queue set being represented by a decision table. Thus, if we are attempting to simulate the behavior of three queue levels and each level contains a specific number of service counters and queues, we would develop three decision tables, each simulating the actions at one queue level. The three decision tables can be coded as three separate subroutines, letting the main program be the "driver" for the simulation.

Analysis of Simulating with Decision Tables

There is a specific set of steps which must be taken when the time is incremented in the simulating of Queue System 1. These steps can be generalized for all simulations using decision tables. However, only the case of Queue System 1 (Model 1) is described for the sake of simplicity and ease of understanding.

The following terms used in the decision table representing Queue System 1 must first be defined:

K	The real time, at increments of five time units.
NEVEN	A counter indicating the number of items or persons in the system at any point in time.
ITIME	A variable used either to place a departure on a queue or to retrieve a departure time from the queue.
KEY	An indicator which determines along with ITIME when a statistical summary should take place.
KA,KB,KC	Flags.
MTL	The queue.

Statement of the Problem: Study the behavior of a single counter with 100 constant arrivals every five units of time, utilizing one queue for the service counter.

The DO loop (K = 5,800,5) will generate the 100 arrivals indicated in the problem statement. The DO loop has an upper bound of 800, since at the end of 100 arrivals—time 500—there will probably still be some persons in the system and we would like to continue running the simulation to determine at what point in time there are no more persons in the system.

The first condition tests whether or no there is an arrival. The two

```
C
C       ENTER DO LOOP WHICH GENERATES ARRIVALS EVERY 5 UNITS OF TIME
C
        DO 1  K=5,800,5
   2    CONTINUE
```

K.LE.500	Y	Y					Y	Y	N					N
NEVEN.EQ.O	Y	N					Y	N	N					Y
ITIME.LE.K			Y	Y	N	N				Y	Y	N	N	
KEY,EQ.999			Y	N	Y	N				Y	N	Y	N	
KA.EQ.O	Y	Y	N	N	N	N	N	N	Y	N	N	N	N	
KB.EQ.O	Y	Y	Y	Y	Y	Y	N	N	Y	N	N	N	N	
J1=0	X	X	X	X			X	X	X	X	X			
CALL GEN (I,KEY)	X						X	X						
ITIME=K+KEY	X						X							
ITIME=K+MTL (1, 1,NEVEN)								X						
CALL OUTPUT (K)			X							X				
CALL TOP (1,J1,ITIME,KEY)		X	X	X					X	X	X			
CALL PUSH1 (1,J1,ITIME,KEY)	X					X	X	X					X	
CALL PUSH5 (1,0,ITIME,KEY)					X							X		
KA = 1		X							X					
KB = 1					X	X			X					
KC = 0	X						X	X				X	X	
KC = 1		X	X	X	X				X	X	X			
KA = 0							X	X				X	X	
KB = 0							X	X				X	X	
KC = 1														X

```
 C   LAST
        IF (KC) 3,1,2
```

FIG. 9-5

flags KA and KB indicate the beginning of a new arrival,[1] while NEVEN discriminates whether or not there is an item being processed at this particular time. If no item is in the system, i.e., NEVEN = O, then the flag KC would be set equal to zero and a new arrival would be generated. At the end of the decision table there is a conditional test of the flag KC. If KC is positive, the decision table is executed again without the generation of an arrival. If KC is zero, execution branches to statement CONTINUE which is part of the DO loop generating arrivals. If KC is negative, the simulation is finished and the program branches to the output routine after the return to the main program. After an arrival is generated, two paths can be taken in the decision table, depending on NEVEN. If no item is in the system, execution passes to the generation of a processing time for a new arrival. The arrival is then placed "in process" (the queue) and a new arrival is generated as the time is incremented by five units. If an item is "in process" at the generation of an arrival, the system must be updated before the new arrival can be placed in the system. This is achieved by a call to TOP [2] which takes the first entry (earliest departure time) from the queue list (departure time list). The value(s) obtained from this call to TOP satisfies one of decision rules 3-7. If decision rule 3 is satisfied, execution proceeds to the output routine after which the same process is repeated until one of the other decision rules 4-7, is satisfied, i.e., decision rule 3 is not satisfied. If decision rule 4 is satisfied there is a departure from the system and we repeat and call the TOP again. Rules 5 and 6 place the output tag and departure time back onto their respective lists (queues) since time has not reached the point at which they can leave the system. Decision rules 6 and 7 generate the processing time for the new arrival. These two rules are executed only after the general updating of the system has occurred. If the queue is empty, the processing time is just added to the current time to obtain the departure time. If the queue is not empty, the processing time is added on the last departure time in the queue in order to obtain the departure time for the current arrival. These departure times are then placed in the queue and execution branches to the generation of the next arrival. Rule 9 is satisfied when there are no more arrivals, K > 500. However, the clock (DO loop) keeps incrementing the time since we want to continue the

[1] The concept of flagging is of paramount importance in simulating with the use of decision tables, for it enables one to discriminate between decision rules while actual conditions of the simulation remain constant at that point in time.

[2] TOP refers to a routine programmed by the user.

simulation of the system until there are no more items in the system. Rules 10-14 have the same effect on the system as do rules 2-6: however, these decision rules do not generate a processing time since there are no more arrivals at the point in time these rules are satisfied. Rule 15 determines when the simulation is completed. At this point the time, K, will be greater than 500 and the queue tag, NEVEN, will be empty. The flag KC is set to —1 and a branch is made to the appropriate location.

The description of the simulation represented by Fig. 9-1 has been very specific. Other, more complex, simulations would be set up similar to this model. However, if more service counters and more queues are added the decision table will become more complex. A decision table which simulates the behavior of two service counters requires 27 decision rules or approximately double that of the single server model. If queue levels are introduced, as was done for queue system 3, the second and third levels representing three and four service counters respectively require only 33 decision rules to simulate their actions. Increasing the number of levels and the number of service counters would not in any way affect the decision tables. The number of decision rules will remain at 33 unless certain restrictive requirements are added, such as priorities, etc. The models studied in this thesis have been of simplest form, and it still remains to introduce other complexities in order to determine fully what effects these will have on the use of decision tables in simulating.

10

Decision Tables for Management Information Systems

Introduction

Decision tables provide a techinque, a format, and a discipline for the analysis and description of all information processing systems. They identify the data manipulations to be accomplished and logic to be followed in data processing applications. They have been applied to a wide variety of other problems, such as highly complex engineering analyses, but this presentation will deal primarily with the format and construction of decision tables in management information systems applications.

Such tables have actually been in widespread use for a very long time. For example, insurance rate tables for different kinds of policies, different categories of risks, different ages of the insured person, etc., are a form of decision table. Another common example is price lists of raw materials and purchased parts in terms of various dimensions, accuracy of dimension, specification of materials, type of finish, etc.

Significant application of decision tables to data processing or to management information systems is a fairly recent development, however. No doubt some use of them had been made previously, but they began to come into prominence in these fields as late as 1960. Continuing attention has been given to their development and use since that time, but their application is even now confined to a relatively small number of organizations.

This presentation will be dealing with management information processes and will be speaking of decision tables in terms of computer applications. Nevertheless, decision tables are equally useful in describing manual, bookkeeping machines, or tabulating equipment processing.

Decision tables are of value in situations where many characteristics or conditions of variable factors must be determined before the desired result can be obtained. One of the values of decision tables is the ability provided to identify all of these combinations with one another in a complete, concise, and logical manner. Through providing such a format and discipline, decision tables make the analysis of complex situations easier.

Decision tables were developed, and are coming into prominence, because other existing methods of problem definition are inadequate. Written descriptions simply do not convey the exact meanings that complex problems require. Many systems analysts are not adept at writing and are inclined to record a wandering narrative about a problem rather than identify the factors of the problem in a succinct and logical manner.

Flow charts have proven to be inadequate and cumbersome. It is difficult to follow the logic of a problem through a flow chart, and the uninitiated person frequently becomes utterly confused with the methodology. This method of documentation is also very difficult to keep current.

Decision tables offer several advantages in overcoming these difficulties:

1. This method of documentation is easily prepared, changed, and updated.
2. It forces consideration of all possible conditions pertaining to the system being developed or analyzed.
3. Because they present systems logic in a clear and concise manner, decision tables allow the systems analyst to communicate with management or the customer without becoming enmeshed in data processing methodology.
4. Decision tables are equally effective in the analysis of a present system or in the development of a new system, either manual or mechanical.
5. These charts allow the systems analyst to communicate with the programmer without encroaching upon his prerogative to handle data in the manner best suited to the computing machine's configuration.
6. When decision tables are prepared within the confines of a programming language such as COBOL or FORTRAN, programming can be accomplished accurately and efficiently directly from the tables.

(*Editor's Note:* At this point, R. Fife's discussion of the basic format of decision tables has been omitted.)

Inventory Control Problem

To further describe the method of entering information on decision tables, an inventory control ledger posting operation has been chosen as a typical data processing problem.

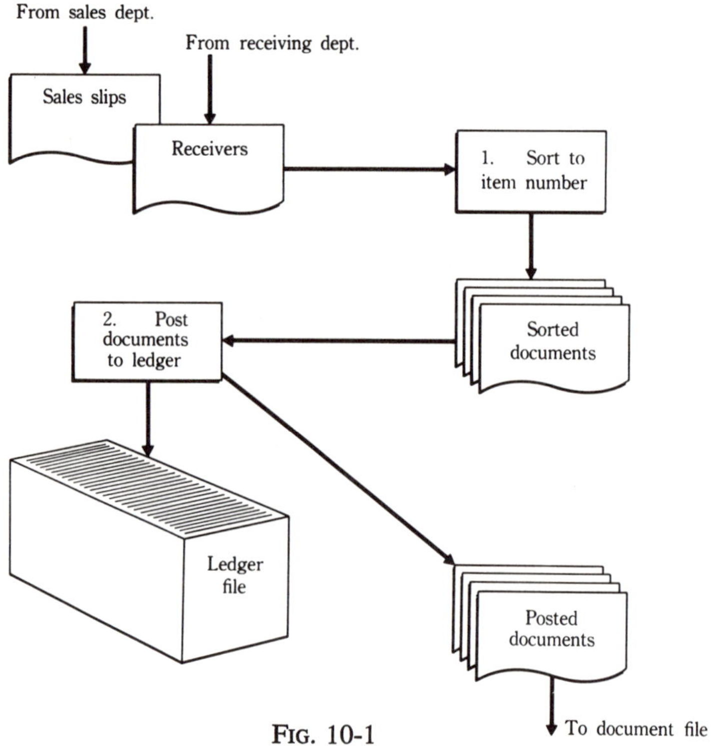

FIG. 10-1

In this hypothetical system, as items are sold, a sales slip is prepared showing the item number and the quantity sold. Also, as items are received, a receiver is prepared showing the same information. These documents are collected and sorted into item number sequence and then posted to a set of ledgers which are maintained in item number sequence. As indicated in the block diagram, the posting of documents to ledgers is a batch processing operation. Each valid document must be posted to a matching ledger. There can be more than one document with the same item number, but there can be only one ledger per item number. In the processing, there can be ledgers which will not match documents. However, if a receipt document item number does not match a ledger, it indicates a new item, and a ledger must be initiated. If the item numbers of the document and ledger items match, receipts are added and sales substracted from the ledger balance on hand. To set up a decision table describing this posting or processing, the conditions which will be encountered must first be entered in the condition stub.

CONDITION STUB
DOCUMENT ITEM # IS EQUAL TO LEDGER ITEM #
DOCUMENT ITEM # IS GREATER THAN LEDGER ITEM #
DOCUMENT IS A SALES SLIP
DOCUMENT IS A RECEIVER

Condition Entries

It is necessary to segregate all of the sets of conditions that could occur within the group of conditions entered in the condition stub. This is done in the condition entry portion of the decision table by filling in a separate column or set for every possible combination of conditions. As the condition stub lines are complete statements, the condition entry sets will consist of "yes's" and "no's" to the questions posed in the condition lines. *Note*: *In the illustration the condition statement "Document Item # is less than Ledger Item #" has not been entered because this condition will exist when the "Equal" and "Greater Than" conditions are both "no."*

It is mandatory that there be at least one "yes" and one "no" for each condition statement. If the condition statement does not apply to the entry set, then the box applying to the statement may be left blank, or a symbol indicating the statement does not apply may be entered.

CONDITION STUB	CONDITION ENTRIES				
	SET 1	SET 2	SET 3	SET 4	SET 5
DOC ITEM # = LEDGER ITEM #	Y	Y	Y		N
DOC ITEM # > LEDGER ITEM #				Y	N
DOC IS A SALES SLIP	Y		N		
DOC IS A RECEIVER		Y	N		

In the example above, five condition sets have been filled covering the following conditions:

Set 1: The document item number equals the ledger item number and the document is a sales slip.

Set 1: The document item number equals the ledger item number, and the document is a receiver.

Set 3: The document item number matches the ledger, but the document is neither a sales slip nor a receiver (this condition is an exception).

Set 4: The document item number is greater than the ledger item number.

Set 5: The document item number is neither equal to nor greater than the ledger item number, therefore it is less than.

Action Stub

Having filled in the condition portion of the decision table, it is now necessary to specify the actions that must be taken for each of the sets of conditions which could be encountered. The action stub of the decision table could contain the following information which is entered in the action stub as illustrated.

Subtract the sale quantities or add the receipt quantities to the balance-on-hand quantities of the appropriate ledgers. Either information pertaining to the sale or receipt and the new balance on the ledger. When the document has been posted, get the next document.

When document item number does not match ledger item number, either get the next ledger or transfer to another table for instruction on the preparation of new ledger sheets.

ACTION STUB
SUBTRACT QUANTITY OF SALE FROM LEDGER BALANCE
ADD QUANTITY OF RECEIPT TO LEDGER BALANCE
ENTER RESULT OF ACTION #1 IN LEDGER BALANCE COLUMN
ENTER "SALE" IN LEDGER TRANSACTION COLUMN
ENTER "RECEIPT" IN LEDGER TRANSACTION COLUMN
ENTER DOCUMENT SERIAL #, DATE & QUANTITY IN LEDGER
GET NEXT DOCUMENT
GET NEXT LEDGER
GO TO TABLE #1 (THIS TABLE)
GO TO TABLE #2 (PREPARE NEW LEDGER TABLE)
GIVE DOCUMENT TO SUPERVISOR
FILE CURRENT LEDGER

Action Entries

The appropriate set of actions to be taken for any condition set is entered in the action column directly below the appropriate condition set column. It should be noted that because all the conditions in the condition set must be true before the set is true, the sequence in which the conditions are tested is of no importance. The sequence in which actions are performed can be of the utmost importance. Therefore, this sequence must be indicated in some manner. In the case illustrated below, a numeral is entered indicating the order in which each action is to be performed. If the action line is not applicable to the condition set, the block in the action set is left blank.

CONDITION STUB	CONDITION ENTRIES				
	SET 1	SET 2	SET 3	SET 4	SET 5
DOC ITEM # = LEDGER ITEM #	Y	Y	Y		N
DOC ITEM # > LEDGER ITEM #				Y	N
DOC IS A SALES SLIP	Y		N		
DOC IS A RECEIVER		Y	N		

ACTION STUB	ACTION ENTRIES				
SUBTRACT SALE QTY FROM LEDGER BAL	1				
ADD RECEIPT QTY TO LEDGER BAL		1			
ENTER RESULT IN LEDGER BAL	2	2			
ENTER "SALE" IN LEDGER TRAN COL	3				
ENTER "RECEIPT" IN LEDGER TRAN COL		3			
ENTER DOC SER #, DATE AND QTY	4	4			
GET NEXT DOCUMENT	5	5	2		
GET NEXT LEDGER				2	
GO TO TABLE #1	6	6	3	3	
GO TO TABLE #2					1
GIVE DOCUMENT TO SUPERVISOR				1	
FILE CURRENT LEDGER			1		

Limited and Extended Entries

LIMITED ENTRIES

Up to this point in our discussion, only limited entries have been shown in the text. In the condition entry portion of the decision table these limited entries consisted of "yes," "no," and "not applicable." In the action entry portion of the table the limited entries consisted of a number indicating the sequence in which actions were to be taken.

EXTENDED ENTRIES

It is very often easier or more normal for an analyst to state a portion of the condition in the entry set. If this is done, the entry is called an extended entry. The decision table illustrated below contains the same conditions and actions of the above table for the ledger posting operation but with extended entries. Note that asterisks are used in the stub statement to indicate extended entries.

	CONDITION ENTRIES				
CONDITION STUB	SET 1	SET 2	SET 3	SET 4	SET 5
DOCUMENT ITEM # * LEDGER ITEM #	=	=	=	>	<
DOCUMENT IS A *	SALE	RECPT	OTHER		
ACTION STUB	ACTION ENTRIES				
COMPUTER LEDGER BAL * DOC QTY	1 MINUS	1 PLUS			
ENTER RESULT IN LEDGER BAL COL	2	2			
ENTER * IN LEDGER TRAN COL	3 SALE	3 RECPT			
ENTER DATE AND QTY IN LEDGER	4	4			
ROUTE DOCUMENT TO SUPERVISOR			1		
FILE CURRENT LEDGER				1	
GET NEXT *	5 DOC	5 DOC	2 DOC	2 LEDG	
GO TO TABLE *	6 1	6 1	3 1	3 1	1 2

The preceding material, by ROBERT C. FIFE, was published under the title "Decision Tables," as part of the *Proceedings of the Univac Users Association*, Spring, 1966. It is used here by permission.

11

Decision Tables for General Purpose Systems Design

One of the large problems faced by a computer installation is maintenance of production programs. Of course, if we were soothsayers we would design the ultimate system the first time around. However, most of us are still looking for the ultimate, and program maintenance is very real. Because of the lack of the soothsayer's foresight, our systems are often short-lived. Sometime, someday, somewhere in the organization someone wants something more.

Many programs and systems are like major appliances—there is some type of short term warranty or guarantee for the user's protection, but after this short term frequently some major breakdown makes repair almost as costly as the initial investment. Even simple maintenance or expansion can cause painful headaches. Certain questions must be considered regardless of how complex or simple is the change to be initiated: at what time within the current system will all the data necessary for the change be present; internal and external control of data must be built in; what type of operational problems may occur in relationship to the change; is there room in the data base necessary for the change—the list at times can be lengthy. Procedural writeups and flowcharts can often answer these questions.

Good flowcharts, documentation, and original design file are some real aids to the maintenance team. If these tools are lacking, the job is comparable to giving a mechanic a wad of bubble gum and a roll of baling wire to repair a Grand Prix racing machine! The results are quite questionable and not at all durable.

Procedural writeups and flowcharts are the origin of basically all systems and must be kept current through all modifications of a system to insure proper handling. A form of documentation that can be useful is decision tables.

A decision table is a form of communication, a method of relating how one should react under certain conditions—may the "one" be man or machine. A decision table may be the link of communication between the manual portion and computer portion of the system, the communication link between systems designer and programmer, and the communication link between programmer and program. The decision table can tell the manual system what must be inputted to the computer systems to obtain the desired results and in what form these results will be realized (reporting structure, variation in data base, etc). The decision table may be used to relate simple, as well as intricate, processes to the programmer. A programmer can then use internal decision tables to communicate with routines within a program.

In the following example of decision tables I would also like to bring out a general concept in file design. Even though a designer must fulfill the needs of an external system, a prime consideration of a system is file design. In this area we can be somewhat of the aforementioned soothsayer. Two of the major causes of change are the introduction of a new condition or the necessity of expanding fields. This can be whipped by isolating codes or keys and numeric fields in full computer words in the case of the package "A" user. The package "B" user does not have the same problems of changing file design, but if the original design provides for extra digits within fields, the actual program change would be minimal. In the example consider a file with two code fields and two accumulation fields to be printed in detail under varying headings that summarize internally for a grand total. In the first case shown. any expansion could cause complete redesigning and programming. In the second case, some expansion space has been provided for. Using this file design and with the use of decision tables expansion of the accumulation fields will be no major problem. The addition of a control code would just mean adding new entries to the program table (one for control code, one for summarization, one for editing). A simple routine of linkage would match control codes (full word or expanded defined field) against the tables and transfer control to appropriate accumulation routine and editing routine.

I would now like to move a system which we at the Great Northern Railway are putting on the air to handle the figuring of withholding in the payroll system. The Great Northern Railway has on-line facilities in ten states and two Canadian provinces and agencies in about fifteen other states. The United States requires withholding for income tax in two manners, depending on whether an employee is married or single;

railroad retirement, which is comparable to Social Security, and railroad unemployment insurance, which is paid by the railroad only. Most every state has some form of withholding. Although Canada has no provincial withholding, it has its national withholding along with a pension system withholding.

Upon examining the situation, we found that on an average of about every six months, one taxing unit or another was changing its method of determining the amount to be withheld from an employee's gross pay.

Another consideration was the probability that a state government could force the Great Northern to pay employees on a different length pay period than that in which the payroll is now handled. This was fairly simple to solve. Each employee's pay period gross pay will be projected to a yearly figure by multiplying the number of pay periods in one year times the employee's gross. Computations of withholding may now be made on a yearly basis and then reduced back to the pay period level by dividing by the number of pay periods in one year.

With the previous payroll system the constant major program maintenance was very time consuming, difficult, and costly. Analyzation as to where the particular changes belonged, or if the change could damage the remainder of the system, etc., were always hairpulling endeavors. To ease the maintenance problem we envisioned a method to change the basic constants in a taxing unit's withholding formula. The method had all constants identified on a master file that was to be updated and read into memory for use by the various formulae. Then the updating of a file maintained a portion of the program.

However, this method did not lick the problem of a taxing unit's changing its entire method of determining withholding. The next concept was an English oriented programming language, one that could be used by most any clerk, to be placed on a master file and read into memory. The language uses the basic mathematical functions of addition, multiplication, subtraction, and division and the comparison functions of greater than, less than, and equal to. The language also has functions to retain answers and terminate an individual routine and a card continuation function.

The order of the master file is taxing unit, category, and sequence. The taxing unit meaning: national level (United States and Canadian), three digit alphabetic; state level, two digit numeric; and local level (city, county, etc,) three digit numeric. The formula may be usually broken down into four basic categories: a gross adjustment routine, range table, withholding determination routine, and a withholding adjustment routine. The fifth category is one of control. It is the answer to a test computation that is performed on the routines once read into memory. The sequence key is used to order items within a particular category.

The contents of the item or card are data names and constants provided by the user and the language's functions. The constants are predefined as being sixteen numeric digits with six before the decimal point and ten after the decimal point. The data names are of two types: (1) values that originate from the employee master file and gross file, (2) accumulations that the user will make. The data names of the first type are listed at the beginning of the withholding language master file so that memory positions may be assigned them. The language itself is translated to pseudo operation codes for the functions and memory locations for data names and constants that are recognizable by one computation routine for a taxing unit formula.

The language is restricted within certain bounds. The functions may be used only in the following predefined manner, and any divergence will be indicated as an error. The basic sequence for arithmetic processes is: data name or constant, arithmetic function, and data or constant "EQUALS" data name. The "EQUALS" is the function with which one retains answers for later reference. The sequence of the comparison situation is data name or constant, comparison function, data name or constant "GOTO i." If comparison conditions are not met, control is transferred to the line or entry after the "GOTO i." "i" equals the number entries to the portion of the routine to handle the passing of comparison conditions.

The items are validated, translated to pseudo codes, and stored into memory. The location of each of the first of the tax unit entries and each subsequent first new category entry under that taxing unit are placed in an identifier table. At the termination of the file a test computation is made if any changes have been made to the file or upon computer operator request via the console. If there is a discrepancy between the previously mentioned control answer and computation routine answer, the system would be jettisoned after testing any remaining routines.

The actual processing of payroll data then begins. Gross item tax codes are matched against the identifier table (taxing unit code) and control transferred to the computation routine with the arithmetic registers loaded with the locations of the categorized routines. An absence of information in any of the "AR'S" indicates no processing for the particular category.

All of the above processing is governed by one programmed decision table or another (validation, translation, identifier match, calculation, and final outputting of the calculation routines). Now we have a master file updating which maintains a much greater portion of the system.

This chapter, by L. E. LESKINEN, first appeared in the Fall 1966 *Proceedings of the Univac Users Association* under the title "General Purpose Systems Design and Programming Techniques." It is reprinted here by permission.

12

Decision Tables for Systems Analysis, Documentation, and Programming

Decision tables have attracted widespread interest in the relatively short time since the first significant experiments were announced. Experience now indicates that they often have clear-cut advantages over other techniques in three rather diverse areas: systems analysis, documentation, and programming.

This paper cites several successful experiments and suggests a course of action for further exploration of the advantages offered by decision tables.

(*Editor's Note:* The discussion of decision table fundamentals at this point has been omitted.)

Applying Decision Tables

Decision tables seem to offer significant advantages in three areas: analysis, documentation, and programming.

They were originally developed for system analysis, as an alternative to or to be used in conjunction with flow charting. In addition to aiding in effective analysis, decision tables also proved useful for communicating the system logic, thus serving as a documentation technique. Programmers, seeing the form, began using it for developing program logic since the clear structure aided them in the complicated areas of their programs.

A few examples of the use of tables will indicate how successful field experiments have been.

ANALYSIS

Mr. A was faced with a very complex logical problem in one section of a COBOL processor he was working on. He was reluctant to use normal techniques for doing this analysis and programming job because of its complexity. He felt he needed a powerful organizational tool in order to see clearly the numerous alternatives. He had no extensive experience with the use of decision tables, but he had attended a one-day seminar on the subject, and decided to give it a try.

After some preliminary study he discovered that there were only seven independent conditions in the problem, and that each of them was of a binary nature. He mechanically wrote down all 128 combinations of these seven variables in the form of a table. He then proceeded to discard the combinations that were impossible because of the construction of the language or because they had been handled by previous processing runs. During this process he discovered a number of cases that called for the same actions; these he grouped together by using "not pertinent" entries in the table. For each rule he wrote the actions required and discovered, somewhat to his surprise, that there were only 20 distinct actions required. The completed table had 70 rules and 20 actions; by the way the table had been constructed, he was certain that all relevant combinations had been handled properly. This table provided the basis for a very simple transition to machine code, in a manner that is quite interesting in itself but unfortunately not relevant here. The soundness of the whole approach was completely justified by the speed with which the resulting program was checked out.

The original space estimate for this job had been 1200-1500 positions; the completed program took 530. Analysis and machine coding required four days, and the job was checked out in just one day.

As an analysis technique decision tables are often more manageable than flow charts or narrative descriptions. They present logical alternatives in an easily-understood graphical form. Since the alternatives are so clearly displayed, they are easy to check for completeness and consistency, leading to thoroughness and accuracy.

DOCUMENTATION

Mr. B had just been informed that one of his programmers was to be promoted and would be leaving within a week. The programmer had just finished his seventh program for their 1401; none of these programs had yet been documented to the point where another programmer could take over. While Mr. B first considered asking the programmer to prepare the usual flow charts, he rejected this possibility because he estimated the job would take three weeks. He had just heard a talk on decision tables and while he had no experience with them he wondered whether they might not provide a solution to his problem. He spent one hour with the programmer, instructing him in decision table techniques and selecting a particular table format. One of the developers of the decision table concept spent a half hour with them to check out their approach, which seemed good.

The programmer then went to work. In just *two days* he wrote the decision tables necessary to display the logic of all seven programs. Since then, the programs have required modification and correction. The programmer taking over, using just the decision tables in conjunction with the machine listings has had no trouble making the changes.

To indicate how effectively a decision table can document system logic, consider the common file maintenance problem: processing a detail file against a master file, both in sequence on identification number. Special actions are required at the beginning and end of each file, and the various combinations of high, low, or equal must be taken into account. Figure 12-1 is a decision table showing the logic of this problem.

Rule 1 states the starting condition. Rule 2 handles the end of job conditions. Rule 3 describes the situation when the end of detail has been reached, but not the end of master. Since there can be no further changes, additions, or deletions to the original master, the actions are to write the updated master from the master area, read another master, and then return to the beginning of the table.

In Rule 4, the end of master has been found, but not the end of detail. Rules 5, 6 and 7 are concerned with cases where neither the detail nor the master file has ended. Rule 5 considers the event when the detail is less then the master. In Rule 6 the detail is greater than the master. Rule 7 covers the case where master and detail are equal.

The final rule, Rule 8, is the "Else" situation. When this occurs something has gone wrong, since all legitimate possibilities have already been examined. Rule 8 will take care of cases involving sequence errors in the master file and certain types of sequence errors in the detail file.

This example shows clearly how a decision table may be valuable for system documentation. Decision logic cannot be presented as con-

	Rule #							
TABLE: Update	01	02	03	04	05	06	07	08
Start	Y	N	N	N	N	N	N	Else
End of Detail		Y	Y	N	N	N	N	
End of Master		Y	N	Y	N	N	N	
Detail vs. Master					<	>	=	
Detail is an "addition"				Y	Y			
Do Error Routine								x
Move Master to New Master							x	
Move Detail to New Master				x	x			
Set Addition Switch				On	On		Off	
Write Master			x			x		
Read Master	x		x			x		
Read Detail	x			x	x			x
Go to Table	Up-date	End	Up-date	Chg.	Chg.	Up-date	Chg.	Up-date

FIG. 12-1

cisely with a block diagram. Drawing a block diagram would take longer than developing this table. There is a much higher probability that the block diagram would contain logic errors and omissions. Decision tables can be read by other people with a minimum of training and explanation.

PROGRAMMING

Customer C has experimented with tables extensively in many areas of data processing. For purposes of comparing tabular-form programming with conventional methods, two similar problems were prepared by equally experienced programmers. Customer C provided the table comparison shown in Table 12-2.

Mr. D, the manager of large scientific data processing center, after careful investigation has instructed his computer people to use decision tables as their basic programming methods.

Mr. D indicates that they have completed twenty-five 704 applications and are extremely enthusiastic about the savings in time that have resulted. They are getting 40 to 80 checked out program steps per hour.

TABLE 12-1. COMPARISON OF TIMES FOR PROGRAMMING TWO EQUIVALENT JOBS USING STANDARD PROGRAMMING METHODS AND TABULAR PROGRAMMING

	Autocoder-level programming (hours)	Tabular-form programming (hours)
Learning	40	40
Analysis	80	21
Flow Charting	62	44
Coding	136	39
Testing	174	54
Total	492	198

As a programming method, decision tables used with a suitable language, reduce program definition, coding, and debugging time. By rearranging the rules in a table, efficient programs can be produced. Tables also provide a natural method of segmenting the program leading to easier debugging and maintenance.

Summary

There has been much progress made in the short time since decision tables were first publicized. Widespread field experiments have demonstrated their strength in analysis, documentation, and programming.

We feel that decision tables deserve even wider study and application, and that such work will lead to significant improvements in the power and applicability of the technique. In particular, we would like to emphasize that the *structure* of a decision table is distinct from the *language* used within it. Any suitable language can be used. Furthermore, the table structure concept is distinct from the characteristics of any programming system of which it may be a part.

We feel that the time is ripe for full-scale investigation of how decision tables can best be combined with other techniques used in the data processing community. We do not propose tables as the replacement for all other systems, but we do feel that tables can be used with other concepts—perhaps in ways not now envisioned—to produce a combination much more powerful than any single part.

This material, by THOMAS B. GLANS, first appeared under the title "Progress in Decision Table Applications," in the Systems and Procedures Association's publication, *Ideas for Management—1963*. It is used here by permission.

13

Decision Tables for Effective Procedures

"It is not enough to be able to write so that you can be understood. You must write so that you cannot be misunderstood." With this philosophy, many organizations are abandoning the traditional narrative procedures in favor of decision logic tables. Decision tables force the breaking of the procedure into the smallest logical components. The conditions upon which actions are based are identified and logically organized to depict every pertinent cause and effect relationship. Consequently, the procedure will more completely cover the situations that might arise in day to day operations.

The decision table format allows a person to read only the area of a procedure with which he is concerned at the moment; he is not forced to wade through unrelated material.

Actions are not only spelled out in a decision logic table; they are also sequenced. This assures the orderly completion of all tasks in the procedure.

Procedures in decision table form, in addition to making the day to day operation run smoother, provide an excellent method of training new employees for the job. They can work with a greater degree of confidence right from the beginning.

The following tables are actual examples of decision tables being used for procedures. The first two are from the accounting department of a business firm. The third is an Air Force procedure for shipping controlled test items.

HOW TO PACKAGE AFPTs DESIGNATED CONTROLLED ITEM (TEST MATERIAL) FOR SHIPMENT

LINE	IF PACKAGE IS	RULES 1	2	3	4
A	to contain complete AFPTs 850	Yes			
B	a carton		Yes		
C	a bundle			Yes	
D	an envelope				Yes
E	include a shipping receipt listing contents (see note 1).		X	X	X
F	group in bundles no larger than 8″ long, 6½″ wide, and 4″ high, so that all cards remain flat. Fold related AFPTs 237 to same size as AFPTs 850 and place on inside top of bundle.	X			
G	insert material in 9″ by 12″ heavy kraft envelope and seal.				X
H	place heavy chipboard or corrugated fiberboard sheets on all sides to protect all sides and edges. Securely tie packages in both directions with a heavy duty mailing twine. Take care not to bend or mutilate contents. Wrap in double sheets of kraft paper.	X	X	X	
I	seal with gummed kraft tape of sufficient strength and width to insure full protection of open seams and ends. Stamp or mark plainly on top and bottom and both ends of inner package: "CONTROLLED ITEM (Test Material). DO NOT OPEN. FOR TEST CONTROL OFFICER ONLY." Place stamping over tape seal so that any attempt to compromise contents will be exposed.	X	X	X	X
J	affix to package a label reading as follows: "FOR TEST CONTROL OFFICER ONLY. THIS PACKAGE CONTAINS AFPRT NR. _____ COPIES NUMBERED _____ THROUGH _____."		X	X	X
K	insert 9″ by 12″ kraft envelope into 10″ by 15″ heavy kraft envelope and seal.				X
L	place package in corrugated and cardboard carton, carefully padding to insure that cards are not damaged. Gross weight must not exceed 10 pounds.	X			
M	insert package in carton. If gross weight does not				

	exceed 30 pounds, use cartons with a minimum bursting strength of 200 pounds. If gross weight exceeds 30 pounds, use cartons with a minimum bursting strength of 275 pounds. Use top and bottom pads of heavy corrugated fiberboard in cartons. Add open cell pads or thicknesses of corrugated board to insure stability when packages do not fit cartons snugly.		X		
N	securely tie twice, in both directions, with a heavy duty mailing twine. Wrap a second time in double kraft paper.			X	
O	seal with heavy kraft tape. Affix mailing labels (see notes 2 and 3).	X	X	X	X
P	band with wire or nylon tape if available.	X	X		

HOW TO READ THIS TABLE:
For example, if the package you are preparing is a carton (line B), then take the actions marked "X" under rule 2.

Notes:

1. Shipping receipts will not be used in transmitting completed answer sheets or AFTPs 850 to the 6570th Personnel Research Laboratory for scoring or disposition.
2. Packages of AFPT materials will be addressed to the attention of the Test Control Officer. Under no circumstances will shipping or mailing labels (including postal registry labels or outside wrappings of a carton, bundle, or envelope) indicate that the package contains AFPT materials. The sender may include numerical codes for his own information if he desires.
3. When completed AFPTs 850 and 237 are packaged, they will be addressed to 6570 PRL (PRPT), CMU Box 87, Lackland AFB, Tex 78236.

ACCOUNTING SERVICES DEPARTMENT
DETERMINING BILLING INFORMATION

EFFECTIVE 06/17/68

RULE ENTRIES

	CONDITIONS	1	2	3	4	5	6	7	8	9	10	11	12	13	14	15	16	17	18
1	IS BILLING TO BE BY SUB?	Y	Y	Y	Y	Y	Y	N	N	N	N	N	N	N	N	N	N	N	N
2	IS BILLING BY DIV. OR DIVISION & BRANCH?	—	—	—	—	—	—	Y	Y	Y	Y	Y	Y	N	N	N	N	N	N
3	ARE CHARGES TO BE CREDITED?	Y	Y	Y	N	N	N	Y	Y	Y	N	N	N	Y	Y	Y	N	N	N
4	IS THE TRANSACTION CODE 904 OR 906?	Y	Y	N	Y	Y	N	Y	Y	N	Y	Y	N	Y	Y	N	Y	Y	N
5	WAS THE MERCHANDISE DELIVERED IN MARYLAND?	Y	N	—	Y	N	—	Y	N	—	Y	N	—	Y	N	—	Y	N	—

	ACTIONS	1	2	3	4	5	6	7	8	9	10	11	12	13	14	15	16	17	18
1	CODE CLIENT & SUB (BOXES 2 & 3). CODE UNIT NO. (BOX 4) IF REQUIRED.	1	1	1	1	1	1	1	1	1	1	1	1	.	.	.	.	.	.
2	CODE DIVISION OR DIVISION & BRANCH AS NECESSARY (BOXES 31 & 32)	.	.	.	.	.	.	2	2	2	2	2	2	.	.	.	.	.	.
3	CODE PAYEE TYPE (BOX 26) A "D" AND COMPLETE VENDOR NO. (BOX 27) OR SECTION 28 IF NUMBER UNKNOWN. COMPLETE CLIENT, SUB, & UNIT NO. (BOXES 2, 3, & 4) IF REQUIRED.	.	.	.	.	.	.	.	.	.	.	.	.	1	1	1	1	1	1
4	CHECK DEL. IN MD. BOX 33	2	.	.	2	.	.	3	.	.	3	.	.	2	.	.	2	.	.
5	CHECK CREDIT BOX 34	3	2	2	.	.	.	4	3	3	.	.	.	3	2	2	.	.	.
6	SIGN & DATE PREPARED BY (LINES 46 & 47)	4	3	3	3	2	2	5	4	4	4	3	3	4	3	3	3	2	2
7	HAVE APPROVED BY MANAGER, ACCOUNTING SERVICES IF CREDIT REQUEST EXCEEDS $300.	5	4	4	.	.	.	6	5	5	.	.	.	5	4	4	.	.	.
8	BATCH PER PROCEDURE 10-10	6	5	5	4	3	3	7	6	6	5	4	4	6	5	5	5	4	4

ACCOUNTING SERVICES DEPARTMENT
TRANSACTION CODE SELECTION

EFFECTIVE 06/15/68

RULE ENTRIES

	CONDITIONS	1	2	3	4	5	6	7	8	9	10	11
1	IS BILLING FOR MANAGEMENT FEE CHNG (CODE 900)?	Y	N	N	N	N	N	N	N	N	N	N
2	IS BILLING FOR EXPENSE CONTROL (CODE 901)?	—	Y	N	N	N	N	N	N	N	N	N
3	IS BILLING FOR RENTAL TAX (CODE 902)?	—	—	Y	N	N	N	N	N	N	N	N
4	IS BILLING FOR FEDERAL HIGHWAY USE TAX (903)?	—	—	—	Y	N	N	N	N	N	N	N
5	IS BILLING FOR MISCELLANEOUS ITEMS (904)?	—	—	—	—	Y	N	N	N	N	N	N
6	IS BILLING FOR TRAVEL EXPENSE (CODE 905)?	—	—	—	—	—	Y	N	N	N	N	N
7	IS BILLING FOR RULES OF THE ROAD (906)?	—	—	—	—	—		Y	N	N	N	N
8	IS THIS A MISCELLANEOUS BILLING (907)?	—	—	—	—	—	—	—	Y	N	N	N
9	IS BILLING FOR PERSONAL PROPERTY TAX (908)?	—	—	—	—	—	—	—	—	Y	N	N
10	IS BILLING FOR ANY OTHER TAX (909)?	—	—	—	—	—	—	—	—	—	Y	N

	ACTIONS	1	2	3	4	5	6	7	8	9	10	11
1	INSERT THE CORRECT TRANSACTION CODE IN BOX 35 AS NECESSARY	1	1	1	1	1	1	1	1	1	1	.
2	INSERT THE # OF ITEMS IN BOX 36 (1 TO 9999)	2	2	2	2	2	2	2	.	2	2	.
3	INSERT THE COST PER ITEM IN BOX 37 ($ & ¢)	3	3	3	3	3	3	3	.	3	3	.
4	INSERT DISCOUNT (IF ANY) IN BOX 38	.	.	.	.	4	.	4	.	.	.	.
5	PRINT MESSAGE (IF NECESSARY) ON LINE 39 (1 TO 4 LINES OF 30 CHARACTERS & SPACES EACH)	4	4	4	4	5	4	5	2	4	4	.
6	COMPLETE BOXES 40 THRU 45 AS NECESSARY ($ & ¢)	.	.	.	.	.	.	.	3	.	.	.
7	COMPLETE FORM USING DECISION TABLE 0106.	5	5	5	5	6	5	6	4	5	5	.
8	RECHECK—SUPPLEMENT BILL MUST FIT IN ONE OF THESE CATEGORIES.	.	.	.	.	.	.	.	.	.	.	X

14

Decision Tables as a Programming Tool

There are four main ways to convert decision tables into computer programs:

- Manual coding from the tables.
- Use of an interpretive computer program that interprets the tables.
- Use of a pre-processor that converts tables into source language for input to an existing compiler.
- Development of a compiler that translates tables directly into machine language.

Of the four methods, manual coding and the use of a pre-processor are the two most widely used at present. Interpretive programs consume a goodly amount of computer time and have not been widely used for this purpose; they might be more widely used if tables were entered into on-line systems. Compilers represent perhaps the most efficient means of converting tables to machine language but they also require more expensive development than do pre-processors for existing compilers. In the early days of using decision tables, it had been desirable to gain experience with tables before developing a table compiler. Hence the use of manual coding and pre-processors.

Also, most of the work to date—for using tables for computer input—has been with Limited Entry tables, rather than Extended Entry tables. Even though Extended Entry tables have desirable features, we would suggest that you start with the Limited Entry tables at first.

Manual Coding from Tables

Before coding begins, it is desirable (and usually necessary) for the programmer to analyze the decision table for completeness, lack of redundancy, and lack of contradiction. Assuming that the tables are of the limited entry type, the programmer should "explode" the table. If there are *n* condition variables, then exactly 2^n simple decision rules should be accounted for, by replacing all dashes with both Yes and No values.

The next step is to decide whether the table will be tested by scanning or by condition testing. With scanning, each rule is tested separately. If the condition entries are set up in binary form, the tests can be performed rapidly and effectively. With scanning, it is no problem to test the ELSE rule last. The most frequently used rules would be tested first.

Most of the effort in the field, however, has been in the direction of condition testing. With this approach, the conditions and the rules are arranged in a sequence that will facilitate testing. The first condition variable is then tested by the program; the results branch the program to one of two tests on the second variable. This testing continues until the program is directed to the action variables.

If the condition testing approach is followed, some means will be desired to pick a good sequence for testing the conditions, so that the number of tests performed is reduced (hopefully minimized) and/or storage space for the program reduced. There are two methods, one for minimizing execution time for the tests and the other for minimizing program storage space. One method is based on minimizing a "dash count" or "weighted dash count" where the dashes represent indifference in condition testing.

The simplest method that we have seen for organizing conditions and rules requires that the conditions be arranged in descending order from most general to least general. That is, the condition with the most Y's and N's (and least blanks or dashes) is placed first, the next most general is second, and so on. When the conditions have been arranged in this sequence, the rules are then sorted. A card sorter has been used for this purpose; a computer sort or hand sort could be used. Blanks sort before N's, which sort before Y's. The procedure does not in general minimize the number of executed tests, but it does provide a fairly effective and simple scheme. Richfield has been using this procedure and reports that, while programmers can develop better programs for small tables, the procedure gives better results than the programmers usually achieve for larger tables.

We are aware of some other work, as yet unreported, that uses the frequency of rule usage and cost of condition tests for obtaining a minimum cost test sequence. A computer would be desirable for performing this analysis for most tables—and could perform the analysis as a part of the translation procedure.

While most of the effort at present is aimed at finding optimal methods for performing condition tests, we wonder if the development of associative memories might not override this work. That is, if the condition entries were stored in an associative memory, perhaps in binary form, the selection of the appropriate rule could be done directly by the memory.

As mentioned, attention has been paid to the frequency with which rules are used (or their probability of usage) in order to minimize the number of executed tests. The programmer might consider putting "rule counters" in the program to keep track of the number of times each rule is used, at least on a sample basis. This information can be used not only to minimize the number of executed tests; it can also be used for optimizing the action routines. The actions associated with the most used rules can be coded for minimum execution time. In fact, file structures might be changed to facilitate the frequently-performed actions.

With condition testing, for minimizing execution time or storage space, it is necessary to be able to change the sequence of the conditions (which condition is tested first) and the sequence of the rules. But actions usually must be performed in the sequence defined by the person writing the table; the sequence may not be changed. Usually this sequence is defined by the order in which the actions are written. By entering a number in the table, one indicates not only that the action must be performed but also its sequence within the set of actions for that rule. If sequence is indifferent for a series of actions, they can be assigned the same sequence number.

There are two types of rules that necessarily require special attention. One is the ELSE rule; as previously mentioned, the ELSE rule must be executed last. If a table is scanned, it is an easy matter to test the ELSE rule last. But if condition testing is followed, it is not so simple a matter. ELSE is tested last with condition testing. Another type of rule might be called the "Initialization" or "first time through" rule; this rule would always be tested first. Again, with scanning, there is no problem. With condition testing, the program for such a rule would have to be split off from the program for handling the remainder of the table.

Program segmentation seems to follow quite naturally from the use of tables. A program breaks down into (a) tables and (b) subroutines whose execution is controlled by the tables. The actions called for in a table can be other tables, formulae, or other straight line coding. This segmentation facilitates program changing.

Also, tables can be used recursively—that is, an action refers the program back into the entry point for the same table. An example might be the setting of a line counter (for printed lines on a page), and then testing to see if the maximum count has been reached. Since condition testing and actions must be clearly separated, recursion is used. The line counter is incremented by an action; the next action is then GO TO SAME TABLE—and the program reenters the condition portion of the table. Note the use of SAME to refer to the same table; a friend of ours has proposed this to avoid the need to refer to the table by name. If the table name is used, and if names are changed, then the maintenance problem is increased.

Use of a Pre-processor

A number of organizations have written their own pre-processors for converting decision tables into source language statements for a compiler. Usually the source language has been COBOL; in one case that we have heard of it has been FORTRAN. Some of these organizations are the Bureau of the Census, the Insurance Company of North America, and the RAND Corporation.

Now, however, organizations that have access to a COBOL compiler may find that they can easily use DETAB/65, a decision table pre-processor written in COBOL. DETAB/65 was developed by the DETAB Working Group of the ACM (Los Angeles). The pre-processor was written in COBOL-61 Required, so that it should be adaptable to most COBOL compilers. Currently, it has been checked out on a number of computers, including the CDC 1604-A, 3400 and 3600 computers, and the IBM 7040, 7044, 7090, and 7094 computers. Interested users can obtain the pre-processor program—in card deck form—and instructions by contacting their computer users group.

With the arrival of pre-processors, such as DETAB/65,, we would expect that the use of decision tables will grow quite rapidly.

Adapted with permission from *EDP Analyzer*, May 1966, published by Canning Publications, Inc., 134 So. Escondido Ave., Vista, Calif.

15

Decision Tables as the Basis of a Programming Language

Decision Tables are an excellent notation for expressing the logic required in the definition of business data processing problems. This definition orientation gives them a distinct advantage over flow charts in specifying the requirements of the problem. Since Decision Tables are an effective problem definition notation which must be translated into computer programs, why not use the computer itself for this translation? The resultant Decision Table based source programming language is characterized by a strong orientation to the problem as opposed to the solution. The procedural solution is the product of compilation.

In recent years several programs have been constructed to translate Decision Tables into some other source language. This pre-compiler approach has been expedient, but may impose restrictions as compromises are required between the Decision Table notation and the basic source language which is supporting it. Some degree of efficiency, flexibility, and clarity may be sacrificed originally. However, the advantages overpower minor disadvantages, and overall efficiency, flexibility, and clarity can be vastly improved.

In 1960 several individuals at the Insurance Company of North America investigated the possibility of using Decision Tables in the development of a large file maintenance program. They constructed a special purpose Decision Table translation program to produce Autocoder statements for the IBM 705. This experiment was highly successful, and their program became the nucleus of a powerful pre-compiler which has been under continuous development at INA since then. The associated Decision Table programming language which is dependent on Autocoder, is called LOBOC (LOgical Business Oriented Coding). That's easy to remember because it is simply COBOL spelled backward. LOBOC accepts coded input and produces English language documentation statements along with the Autocoder program. LOBOC development was originally under the direction of Lynn Brown, who was at the time a member of the Data Transformation Task Force of the CODASYL Systems Group, along with Sol Pollack and Burt Grad, who are also on the panel today. By 1963 LOBOC had become the principal programming language for IBM 7080 projects at INA. Now INA has eleven System/360s on order, and eight programmers working on a major extension to the 7080 LOBOC pre-compiler to permit it to also accept as input a combination of System/360 and Decision Table statements and produce source language programs for the System/360.

What are the necessary components of a Decision Table programming language? There are four syntactical subdivisions, which are discussed individually below.

Data Division

The data division shares the requirements of common data processing source languages, with some additions. Data conditions must also be defined. Data names are assigned to the various data entities and condition names are assigned to the significant conditions of these data entries. Imposing this facility on an existing source language may cause some loss of clarity due to the added complexity. The skilled programmer quickly overcomes this.

A task group of the CODASYL Systems Group is currently considering some problems in the development of a machine-independent, definition-oriented data division. Their conclusions may be helpful in the evolution of the data division for a Decision Table based programming language.

Action Division

This division consists of action routines. The conditions under which they will be executed are indicated in the decision division. These routines are written in a macro-level language, and ideally are completely free of explicit condition testing. Historically Decision Table pre-compilers have been gradually developed, and the adopted action language has had a wide range of expression. Consequently provision often exists for condition testing in the action routines. Explicit condition testing should be excluded wherever possible. Implicit condition testing will, of course, continue to be required in a sophisticated macro language. For example, the use of a macro to search a storage file implies that condition testing coding will be generated in the course of compilation.

Decision Division

This division consists of Decision Tables exclusively. These tables should be extended entry, vertical rule tables with the following two restrictions. Within any given table, one and only one rule may be satisfied. Disjointed sets of conditions may not be tested by different rules within one table, even when these rules state mutually exclusive conditions and are overtly logical. This is to say that in the generated coding, once the conditions specified in a set of rules begin to be satisfied and further inspection proves that all of the conditions of none of these rules are completely satisfied, it is assumed that no rule is satisfied for this table. No linkage is generated to test for the possible satisfaction of other rules. These restrictions insure against exaggerated complexity in a single table.

LOBOC has used a horizontal rule format. Each condition test required two operands. These operand lengths were set to two positions so that all of the conditions for one rule might be punched on a single card.

These operands specify data names. This limited data names to two digits, allowing a maximum of 1,496 unique alphanumeric data names for an entire program. This also restricts the range of expression in data names. To remove this limitation, a major revision to the LOBOC precompiler will convert the Decision Tables to a vertical rule format. Aside from increasing the number of usable data names, a vertical rule format improves the readability of the unpunched coding form.

Although these are nonessentials, Decision Table notation should provide the following facilities:

1. The specification of indexes to modify designated operands of a table at object time. In addition to improving general flexibility, this facility would permit the Decision Table language to be used in constructing general purpose programs, such as sorts.

2. Indication of the housekeeping routines, if any, which are associated with each table as a part of the notation of that table itself. These

routines might be categorized as first time only operations, execute every time operations, etc.

3. Simplified updating procedures, to permit updating of single items and sets of items selectively.

Director

The director provides the pre-compiler with the following information:

1. The order in which the components of the program are to be presented to the compiler.
2. The rules for segmenting the program into overlays, or separate machine runs.
3. Environmental data.
4. Miscellaneous instructions. These include instructions to make classes of alterations to the entries on the history file, which are used in pre-compile operations.

Essentially the programmer uses the director to order the individual components of the program and to communicate special instructions to the pre-compiler. This technique provides great flexibility for major revisions to the program layout. The director can also be organized to compile only selected portions of a series of programs.

Central Linkage Control

Central linkage control is a system subroutine which is automatically inserted in the object program to supervise the execution of the generated routines. The action strings in the action section of the Decision Tables are the only procedural notations in the language. The flow path of the object program is generated from these strings. The action section may contain three classes of statements: the name of an action routine, the designation of a single common operation, or a reference to a Decision Table.

Central linkage control is not involved in the first two classes. The operation or action is either generated in-line, or linkage is provided for its execution as a subroutine. This is effective because the coding in an action routine may not branch to another action routine or to a Decision Table. Thus the return linkages remain simple.

Three types of reference to a Decision Table may occur in the action strings: DO, GO TO, and GO TO NR. These are the principal procedure notations of the language. These operations work with central linkage control and direct linkage and branching between tables.

DO generates a return linkage address, passes it to central linkage control for filing, and branches to the specified table. Central linkage control keeps track of the return linkage addresses in the order in which they are received by storing them in a push-down list. Return linkages are executed in the reverse of this order. The last one stored is the first one executed. This is accomplished through the generated coding in the following manner.

A branch is generated to central linkage control when an action string terminates without indicating what is to be executed next. This notation indicates that a return is to be made on the last stored return linkage address. Central linkage control executes a return linkage for the last recorded address, and erases its recollection of that linkage. Thus a series of DO operations may be performed without interruption by a return operation and vice versa. This technique permits great procedural complexity, which has been simply stated in the action strings of the Decision Tables, to be handled without difficulty.

GO TO statements generate a simple branch to the specified table. Central linkage control is not affected.

GO TO NR statements generate a branch to the specified table and wipe out all return linkage addresses stored in the push-down list of central linkage control. NR stands for No Return.

Each Decision Table is generated with a single entrance and a multitude of exits through its various action strings. All branches to a table reference use this single logical entry point. Branches to a table may be specified by only one of the three statements listed above.

The use of central linkage control subroutine permits the object program to share the basic modularity of the source program. This permits revision to the basic flow patterns at object time and simplifies patching in general. These features tend to diminish the time and effort required in program testing and debugging.

Conclusion

What are the advantages of a Decision Table programming language?

1. *Flexibility.* The source language is definition oriented. Changes in the source language, reflecting an alteration in the problem definition, automatically produce a revised procedural solution which may differ a great deal from the previously developed solution. The programs are built with modularity. Major organizational revisions can be made by a few entries in the director. Each action routine occurs only once in the source language, although it may be inserted numerous times in the generated program.

2. *Clearness.* Decision Tables are automatically documented with a series of English language, table format statements at each compilation. This removes the burden of maintaining documentation from the programmer. The modular construction of the programs also makes them more easily understood.

These two factors simplify the process of turning a program over to a maintenance staff and generally reduce the cost of implementing program revisions.

Although a Decision Table programming language is still a procedural language, it is also problem oriented and powerful. A few statements represent a lot of logic in an easily understood and revised form. Its use simplifies the communications between the system analyst and the programmer. It even has the potential of removing the distinction between the two. Although individual compile times increase when using a Decision Table language, the overall time and effort expended in problem defining, programming, and debugging is reduced. This savings becomes more significant as the size of the programs increases.

A Decision Table language is not fashioned after the English language. Its notation is borrowed from symbolic logic and it is developed specifically for solving data processing problems. The language is designed to make simple, clear, unambiguous statements. The fact that Decision Table languages in use today depend on Autocoder or some other machine oriented source language is a result of their gradual development. The characteristics of Decision Tables make them well suited to be the basis of a new machine independent source language which would successfully compete with any of the high level source languages in existence or under consideration today.

This chapter by DONALD DEVINE, appeared in *Data Processing, Vol. VIII* (the Proceedings of the 1965 International Data Processing Conference). It is reprinted here with permission of the Data Processing Management Association, Park Ridge, Ill.

16

Decision Tables as an Extension to Programming Languages

The many computer languages available to programmers may be divided into four groups in increasing order of complexity:

1. Machine Languages
2. Symbolic Assembly Languages (e.g., SOAP)
3. Macro Languages (e.g., AUTOCODER)
4. Compiler Languages (e.g., FORTRAN, COBOL)

For most programmers, levels 1 and 2 are of only historical interest. The vast majority of programs written today originate in a macro or compiler language. The current developmental trend in these higher level languages is toward increased sophistication in individual statements without corresponding improvement in clarity or legibility. It is common to discover reactionary programmers wallowing around in the dark ages of FORTRAN II or COBOL 60, while compiler writers go merrily on to later versions. COBOL was at one time highly touted as light reading for executives—the self-documenting language—but in practice it appears only to allow programmers to work in increasingly complex pidgin English.

Many reasons have been suggested to explain why recent improvements in programming languages seem to be ineffective. One possibility is that all the levels of language listed above are sequence oriented, one-dimensional strings of commands. Since computer programs are many dimensional, such languages, however sophisticated their statements, always will be inadequate as program documentation. It is a rare sheet of programming which does not have many branches to other sheets, and continuity of information is lost, rather like trying to read a paper which is all footnotes, or possibly, to read only the front page of a newspaper. In practice, programmers read programs with the help of block diagrams, and these diagrams give them the necessary additional dimensions of complexity. Decision Tables appear to be the only practical method yet proposed to present multi-dimensional information without resorting to graphics.

The purpose of this chapter is to show how Decision Tables may be incorporated into higher level languages. Apart from the previously suggested advantages of better documentation, we have found that they force programs to be logically complete and to be more concise. Further comments on documentation can be made later after the form of the suggested tables has been defined.

Table Structure

In order to simplify machine processing, limited entry tables are proposed. The table structure shown in Fig. 16-1 has proved to be practical.

Table Description	
Table Set-up Optional	
Condition Stub	Condition Entry
Action Stub	Action Entry
Exit Stub	Exit Entry

FIG. 16-1

Since this is a limited entry table, condition entries are YES, NO, or BLANK; action entries are BLANK or NOT BLANK. The statements in the table are written in regular computer language with some ground rules imposed.

Table Language

If Decision Tables are to be used as sub-sets of the total program, they should be allowed to be inserted arbitrarily in the program deck. For assembly and documentation purposes, the table cards should have a distinctive standard identification so that the assembly program can recognize them. Working in AUTOCODER language, we found it convenient to omit page and line numbers from table cards as an identifying code. Working in other languages, such as FORTRAN, a different indicator could be chosen: for example a characteristic punch in column 72.

Distinctive statement labels also are recommended. We used the following label format: TTTYII where TTT are three alphabetic characters identifying the table; Y is C, A, or X as the statement is a condition, action, or exit; and II are sequence digits. Other schemes could be used, but they should be distinctive for ease of assembly.

Table Description. All higher level languages allow for comment or description cards which are non-executable statements. Such statements may be used for the description area, identified as table cards.

Table Set-Up. It often is convenient to be able to enter a table in two places. A preprocessing or set-up phase which will do housekeeping, clear counters, set index registers, etc.; and the regular entry, which is the first condition statement. A set-up entry is particularly valuable if the decision table is an iterative routine. Statements in the set-up area are written in standard format, identified as table cards. The set-up entry should have a distinctive label.

Condition Stub. All higher level languages include a logic statement which is written in the form.

Command, Logic Function, Yes Branch, No Branch

This statement may be truncated by dropping the YES BRANCH and NO BRANCH and used without further change in the condition stub. The assembler actually will assign the branches during a pre-compile assembly, based on condition and action entries. Condition statements should be labeled, using some standard format.

Condition Entry. The last columns of the condition statements are reserved for a truth table, which is the condition entry. The elements of this table are Y, N, and blank. It greatly simplifies the assembly of decision tables if this truth table is required to have a valid tree structure starting at the upper left-hand corner. This requirement is illustrated in Fig. 16-2.

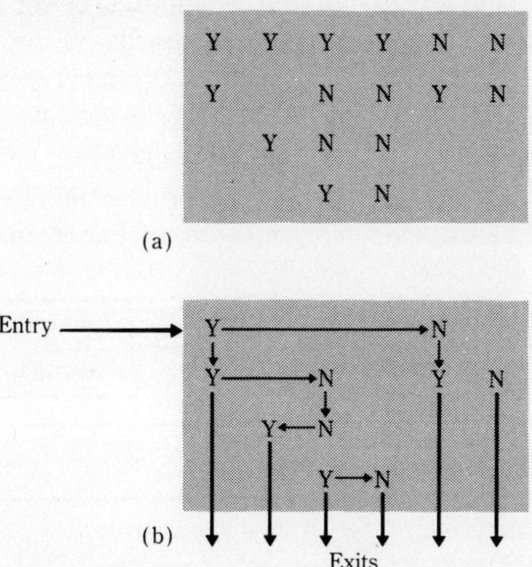

FIG. 16-2 (a) CONDITION ENTRY AS WRITTEN; (b) CORRESPONDING AREA AND BRANCHES

Note that the first row may only have two branches; in general, the *j*-th row may have no more than 2 *j* branches.

The restriction of requiring a tree structure in the truth table is not as restrictive as it may appear, since it follows most programmers' natural thought pattern. Tree structure errors can be readily caught by the assembly program.

Action Stub and Action Entry. On the basis of the conditions in the condition stub area, the condition entries select an appropriate column. Nonzero entries in the action entry area select those actions in the action stub which are to be performed. This is normal decision table logic. The action entries are written in standard compiler or assembly language, with the last card columns reserved for action entries.

At least one refinement to the basic concept is well justified. Action statements should be treated as short strings; only the first member of this string need be labeled, and only this first statement needs an action entry. Subsequent statements in the short string are automatically executed until the next labeled statement is encountered.

Exit Stub and Exit Entry. The exit stub and entry areas really are specialized actions. The only difference is that, in exit areas, operations are restricted to branch commands which leave the table. This separation has two advantages: it makes assembly easier (no point in branching to a branch), and it is an aid in program documentation.

Assembly of Decision Tables

An effective method of converting the programming language decision tables to compiled machine coding is diagrammed in Fig. 16-3.

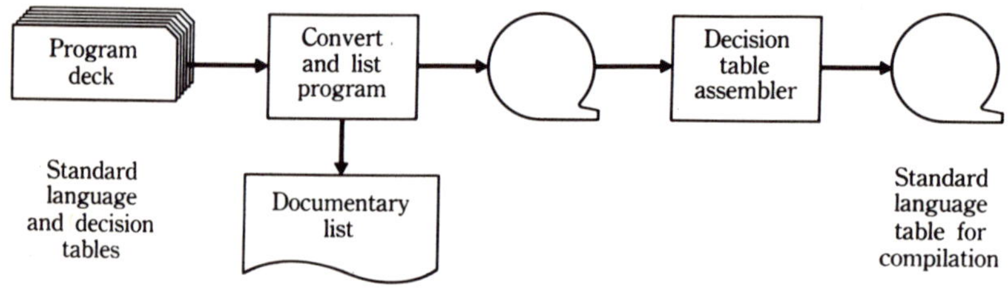

FIG. 16-3

The conversion and listing program may be run on a peripheral computer. The Decision Table assembly may be run as an optional program to immediately precede normal compilation.

It is not the purpose of this chapter to dscribe the Decision Table assembly program, which in any event must vary with the language and machine used. A rough outline of the steps required is:

1. Clean out redundant entries in the condition entry area.
2. Extend the condition stub by adding enough logical statements so that each will have only one YES and NO entry.
2. Create logic statement exits and new housekeeping commands to select proper strings from the action stub corresponding to each YES or NO exit.
4. Merge new statements with old and write assembled programs on output tape.

A good deal of ingenuity can be used to optimize this assembly and eliminate redundant commands. Logical completeness of the table also can be checked during assembly.

Documentation

There is a strong implication in the first section of this report that decision tables eliminate the need for block diagrams. Unless the program is simple enough to be written as a single Decision Table, this is not true. For larger programs, the decision table listings will document detailed logic and actions, and a single comprehensive block diagram showing the interconnections between tables will document the whole program. If the following rules are established in programming, this can be reduced to a standard routine or even mechanized.

1. All branches (except possibly trivial branches, like error conditions) should be made through condition entries in decision tables.
2. The table description should be descriptive enough to completely identify each decision table on the block diagram.
3. All interconnections between tables should be effected by statements in the table exit stubs.

Closed subroutines can, and probably should, be written as decision tables. These will not show interconnections as their return is to an indexed address.

There seems to be reason to believe that decision tables can be used, if not to improve, at least to standardize, in-house program documentation.

This chapter by H. I. Meyer, appeared in *Data Processing, Vol. VIII* (the Proceedings of the 1965 International Data Processing Conference). It is reprinted here with permission of the Data Processing Management Association, Park Ridge, Ill.

Note: Complete logic diagrams which may help in writing a new assembler are available in the reference: "Autocoder Decision Table Assembler," H. I. Meyer, IBM Reference Number 7070-01.1.002.

17

Programming in a Modular Form with Decision Tables

The term "modular programming" as used in this discussion may be defined as the breakdown of the functional processes to be performed in a computer application so as to mentally and within a program separate environmental processes from the procedural or action processes. The mental division allows one to defer certain logic problems until such time as it is necessary to initiate this logic. The program breakdown allows one to perform one individual job in one place. Individual jobs may be identical. If this is the case distinct individual routines access a common routine.

The closed subroutine is the foundation of modular programming. It permits identical tasks to be performed by one routine that may be accessed from different portions of a program,. This means "debug" the routine once even though it may be used by more than one environmental condition. The very nature of the closed sub-routine and the transfer return instruction (TR) provides a "debugging trace" throughout the program.

The processing within a vast majority of programs can be divided into three basic groups: examination of the environment; the linkage between environment and action; and the action processing. The examination of the environment is normally a testing process in which the programmer determines from the data base the type of action processing required. The linkage routine transfers control from the environment processing to the appropriate action processing. The action processing normally handles various updating of the data base and the reporting structure of the program. Each of the three groups would be broken down in as many closed sub-routines as would be necessary to perform the various jobs of testing, linkage, and action processing.

In the examination of environment programmers often test one code and when this code is identified they move to the next code, identify it, and so on through any codes to a final conclusion of which process must be performed. I have a suggestion we may consider at this time. Why don't we examine various control codes as a *group* of codes and identify them as a *group?* When a programmer examines one code at a time his flowchart may depict a simple line of logic. But once in an actual program this logic seems to bounce around like a ping-pong ball, building the final conclusion from several other minor conclusions. The transfer of control or the linkage with the action may be initiated from several different places because of the many different logic paths. Whereas, when examining the codes as a group, the logic of the program will be one consecutive flow from one code group to another. The final analysis is made with one test, not several tests on singular codes. The transfer of control will originate at several different points, but the transfer will be more closely related to the codes.

When maintaining a program that checks the codes one at a time, one must be careful that all tests related to the maintenance are changed and that communication linkage is properly altered. In the program that tests the codes as a group, the maintainer need look only at one test or constant and also able to locate the linkage much easier. In a program using decision tables one need only change the appropriate table and add a routine or alter a routine already listed in the table.

The documentation decision I use as an example (see below) is a portion of a much larger table. The control is made up of two separate codes, an item and a change code. The numbers in the decision squares are to indicate the sequence of processing. If you will notice, three of the control codes use the same edit, but to keep a true modular form there would be three distinct closed subroutines that would merely transfer control (TR) to the common edit routine. In the case of the erorr entry and the updating of the data base the decision table refers one to other decision tables to perform the required logic. This could indicate a deferment of a logic problem.

Using a programmed decision to control the examination of environment and linkage, the group of codes, item type, change code, and transaction code can be matched by going through one function loop and transferring control at *one* point to the routines which follow the control code in the decision table.

```
DECISION  ALPF 2013, : Control Code
          ALPF    1, : Control Code
          LOCA VAL20131, : Validation routine
          LOCA ER20131, : Error routine
          LOCA TR20131, : Transaction routine
          LOCA ED20131, : Edit routine
          ALPH 2037,
          ALPH    3,
          LOCA VAL20373,
          LOCA ER20373,
          LOCA TR20373,
          LOCA ED20131,
```

One could leave the transaction code out of the control code and perform the decisioning and actual processing within each respective edit routine. This would shorten a decision table but not separate action and environment. This could cause problems at time of maintenance for the transaction routines would be buried in the edit routines.

Programming in a modular form has many benefits. Modular programming may shorten program "turn-around time": (1) The built-in trace function can cut down the time spent in "bugshooting." (2) In its strictest sense modular programming provides for the performance of identical tasks in one closed subroutine hence making it necessary to write and "debug" the task once and only once. Modular programming eases any later program expansion or modification. I believe with this in mind all of us should strongly consider modular programming.

TABLE 17-1. MASTER FILE MAINTENANCE

	ACTION												
CONTROL	VALIDATION				ERROR								EXIT MAINT.
					=		≠						
	1	2	3	4	Err. Tab.	Exit Maint.	Appl. Tab 1	Appl. Tab 2	Edit 1	Edit 2	Edit 3		
2013	1				2	3	2		3		4		5
2021		1			2	3		2		3			4
2029			1		2	3		2	3				4
2037				1	2	3	2		3		4		5

This article, by L. E. LESKINEN, originally appeared under the title "Programming in a Modular Form With or Without Decision Tables" in the Fall 1966 *Proceedings of the Univac Users Association.* It is used here with permission.

18

Decision Tables Directly into Programs

The Detab/65 Language

PREFACE

In June, 1963, Work Group 2 of the Special Interest Group on Programming Languages (SIGPLAN) of the Los Angeles Chapter, Association of Computing Machinery, was formed to develop a preprocessor for DETAB-X, a COBOL-oriented decision table language. The results of that effort are partially reflected in this article. Since these specifications differ significantly from those originally set forth for DETAB-X, it is denoted as DETAB/65.

The source document for this article is "Preliminary Specifications for a Decision Table Structured Language-DETAB-X," issued by the CODASYL Systems Group at a symposium on decision tables co-sponsored by CODASYL Systems Group and JUG (Joint Users Group) in September, 1962. Upon these specifications, the SIGPLAN work group has based further efforts to develop detailed specifications. In doing so, these basic deviations from DETAB-X have resulted:

1. DETAB-X prescribed extensive revisions to the structure of the COBOL Data Division, most notably a fixed format for data declaration. In view of the efforts of other groups in developing fixed

Author's note: A preprocessor for decision tables using the COBOL *language was developed in the Summer and Fall of 1965 by a group of interested parties which included the author. Their writeup of this preprocessor is in two parts: (1) A description of the language; and (2) A user's manual.*

To help those who have been voluntarily distributing these specifications, we are making them a part of the "Ideas for Management."

formats for COBOL, this portion of the DETAB-X language specifications was deemed an unnecessary and redundant effort.

2. To conserve space, DETAB-X specified several short forms and substitute expressions (DO rather than PERFORM, SET rather than COMPUTE, etc.) and placed restrictions upon allowable COBOL expressions within decision tables. To maintain maximum compatibility with the COBOL language, placing as few restrictions on the programmer as possible, these short forms and limitations have been removed. Any COBOL expression (legal or illegal) is permissible and may be used at the programmers' discretion.
3. The expression NOT has been added to Extended Entry Condition Entries, meaning all conditions not otherwise specified.
4. A special section of table-specific formulae were specified in DETAB-X. In DETAB/65, formulae too long to be part of the Condition or Action Stubs or Entries are relegated to COBOL code.
5. In DETAB/65, a decision table input to the preprocessor results in a COBOL *Section* being generated. These sections are set up as closed subroutines and must be treated as such by the accompanying COBOL code.
6. Several other changes of a more minor nature are treated in the text that follows.

SIGPLAN (DETAB) Working Group 2 was chaired by Wim Boerdam of Richfield Oil. The principal participants and their affiliations during the development period were:

Mike Callahan, N. E. Willmorth, and Anson Chapman, System Development Corp.; Charles M. N. Cree, IBM; Robert L. Dover, Barry Smith, Control Data Corp.; Stanley Naftaly, Lockheed Aircraft; Solomon L. Pollack, North American Aviation, Inc., S&ID; Wylie Robertson, IBM; Richard W. Senseman, Univac: Ralph Shoffner, Informatics.

Other participants have been George Amerding of RAND Corporation, who advised the group on FORTAB; R. T. Fife (now of Univac); Leonard Longo (then of Douglas); C. J. Shaw of SDC (presently Chairman of SIGPLAN); Charles Powell of Richfield; and Ed Manderfield of North American Aviation and previous Chairman of SIGPLAN.

These specifications and the DETAB/65 processor are being distributed through JUG. Requests for copies of the specifications or for the DETAB/65 processor should be addressed to: Miss Joan Van Horn, Secretary, JUG, Mitre Corp., Bedford, Mass.; or an affiliate of the JUG organization. Correspondence on technical error, comments, criticism and suggestions may be directed to: Mr. Wim Boerdam, Richfield Oil Corp., 645 South Mariposa, Los Angeles, Calif.

Section 18-1. The Detab Language

PURPOSE

The purpose of DETAB/65 is to provide a practical foundation for experimenting with a decision-table based language. The language is designed to be convenient for preparing a preprocessor to go from DETAB/65 to COBOL-61. As such, many restraints and limitations have been placed upon the language to make it readily compatible with COBOL-61.

The DETAB/65 language is designed to fit within the framework of the COBOL language. Decision tables input to the DETAB/65 processor will be output as closed subroutines and treated as COBOL sections. Normal COBOL program formats are used. Symbolic data references made inside decision tables must be declared in the DATA DIVISION just as for non-decision table sections. Formulae whose names are given in a table must be expressed in the PROCEDURES DIVISION prior to entering the DETAB/65 section that gives the formulae values. Within a decision table all normal COBOL expressions may be used, plus a few minor language extensions and symbology necessary to the direction of the DETAB preprocessor and construction of DETAB expressions.

DECISION TABLES

Of the various activities that go into setting up a data processing procedure for a computer, one of the more difficult is the development of a definition of exactly what is to be done under all combinations of circumstances of the data processing problem. Every problem step must be specified. The conditions to normal processing must be identified. Necessary sequences of operations must be precisely indicated.

Determining what is required of the computer system is called analysis; deciding how to go about meeting these requirements is the area of system design; and communicating the chosen procedure to the computer is called programming. In each of these areas a language is needed for defining the data processing procedures. Ideally, a language form or structure should be suitable for man-to-man and man-to-machine communication.

Many languages are used for these purposes. Procedures are often communicated to the machine in a form closely resembling the language of the machine. Symbolic logic and equations are sometimes used, but this imposes a heavy and unnecessary burden on the person writing the procedures. This condition occurs because human language, used for man-to-man communication, and machine language are quite different. Flow charts are widely used for man-to-man communication about data-processing procedures. However, such charts have several drawbacks: Flow charts can become confusing in complex situations; it is relatively difficult to check all possible paths; and the flow chart form is not particularly suitable for direct communications with the machine. Flow charts sometimes present logical equations, but they do not display relationships in as graphical a form as one might wish. Furthermore, they are not a comfortable form of expression for most system designers, except to the person who designed the program.

Decision tables offer the promise of nullifying and correcting many of these language objections. Decision tables provide a graphical representation of complex procedures in a way that is easy to visualize and understand. They show alternatives and exceptions much more explicitly than other language forms. They present relationships among variables clearly. They show the necessary sequences of conditions and actions in an unambiguous manner. Decision-table form can be used with equal effectiveness for system analysis, procedure design, and computer coding. Thus, a computer procedure written as a set of decision tables is, to a large extent, its own documentation.

There is a growing body of evidence to indicate that these claims are justified. Those who have used decision tables for man-to-machine work say that:

1. programming is much faster;
2. program checkout time is significantly reduced;
3. the use of tables leads to greater accuracy and completeness in problem formulation;
4. program maintenance is simpler; and
5. a program written in tabular form is indeed a powerful communication and documentation device.

(*Editor's Note:* S. Pollack's discussion of the structure of decision tables has been omitted at this point.)

ORGANIZATION OF THE MANUAL

The balance of this language specification manual covers the individual areas of the language, indicating the characteristics and restrictions. In general, the rules of COBOL-61 are followed explicitly except in matters of format. Where there are differences, these are noted or are self-evident through the text itself.

Section 18-2. Source Program Format

The format for a COBOL program containing DETAB/65 decision table expressions must conform to the requirements for any COBOL program, except that a decision table may be inserted in the PROCEDURES DIVISION as a SECTION. In compilation a decision table will be treated as a closed subroutine; that is, as a closed COBOL PROCEDURE. Within the table, of course, transfers to other than the normal return point may be specified.

IDENTIFICATION DIVISION

A normal COBOL division name followed optionally by a PROGRAM-ID and other identifying information must be given. No special requirements are levied by DETAB/65.

ENVIRONMENT DIVISION

The Environment Division must be filled out as required by the particular implementation of COBOL-61. Minimum required entries are CONFIGURATION SECTION with SOURCE-COMPUTER, OBJECT COMPUTER and SPECIAL NAMES, INPUT-OUTPUT SECTION with FILE CONTROL and I-O-CONTROL.

DATA DIVISION

The Data Division for a program incorporating DETAB/65 sections is treated as is any other COBOL Data Division. Any symbolic data references used in a decision table must be declared in the Data Division as for any other COBOL procedures section. Any data structures, working storage, and constants used by the program must be described in the Data Division. Symbolic references used within a decision table must conform to the requirements of COBOL data references. Any COBOL data description forms that have been implemented in a particular COBOL processor may be used.

The normal COBOL character set, as implemented in a particular computer, may be used to form NAMES.* One such name will be a DECISION-TABLE-NAME, which is a name given to the procedure table that describes a series of conditions and actions, and which is equivalent to a PROCEDURE-NAME in COBOL-61. A DECISION-TABLE-NAME is a SECTION-NAME and may be composed of alphabetic, numeric, alphanumeric, or combinations of these characters joined by one or more hyphens (-). The DECISION-TABLE-NAME must be used to call a table from the main sequence of COBOL instructions.**

PROCEDURES DIVISION

The Procedures Division of a COBOL source program is used to specify the logical decisions and actions that provide the desired processing. Procedures are normally written as COBOL SENTENCES that are combined to form COBOL PARAGRAPHS, one or more of which may be combined to form a COBOL SECTION.

However, within a section that is a decision table, the normal sentence structure of the COBOL language is abandoned in favor of the more formal structure of the decision table. The syntactic content of the decision table structure may be interpreted as a complex set of conditional statements, plus the information necessary to initialize the closed subroutine that the table represents. A decision table may not be entered via the normal operating sequence and may be referenced by a GO TO, or a PERFORM, but not by an ENTER. A GO TO results in an unconditional or conditional transfer to the specified decision table. For a GO TO from the main processing sequence from

* *Note:* Special symbols suggested by the DETAB-X Manual have been rejected in favor of COBOL forms, i.e., / = (not equal), ◀ = (less than or equal), ▶ = (greater than or equal to) are rejected.

** *Note:* The so-called "short form" of decision table names specified in DETAB-X will not be implemented in DETAB/65. Neither will the capability of calling the TABLE-ID (e.g., GO TO TAB XXX) rather than the table name.

another table, a return or transfer point must be specified by the programmer in the decision table actions or the processing sequence will be lost.

If the transfer to the tables is accomplished by a PERFORM, a normal return to the processing sequence will be made unless the programmer specifies otherwise. Specifically, the DETAB/65 processor will generate a GO TO DEXIT for every rule that does not end in a GO TO, however the table was entered.

Section 18-3. Decision Tables

This section describes the decision table format expected by the DETAB/65 processor. The processor will accept both extended and limited entry tables.

Each section of a decision table will be discussed, each entry permissible in a section will be described and the reasons for the various requirements and rules governing the entries will be given. The description assumes that punched card or card images are used as the input mode. If punched tape, typewriter keyboard, or other continuous input mode were adopted, considerable revision of the DETAB/65 processor Data Division would have to be respecified.

A decision table consists of 6 parts or sections: a Table Header, a Rule Header, a Condition Stub, a Condition Entry Area, an Action Stub, and an Action Entry Area. A sample DETAB/65 specification form is shown in Fig. 18-6 to provide a ready reference for the reader as the decision table sections and entries are discussed.

FORM HEADER

The Form Header serves to identify system, program, and author of completed Decision Table Input Forms, and to specify the data of completion and enumerate the pages. None of this information is part of the decision table proper.

TABLE HEADER

Each table has two header lines: (1) a Table Header that serves to identify the table and provide information that covers the table as a whole; and (2) a Rule Header that is used to indicate rule numbers.

A table may require more than one page, either because of a large number of rules, or because of a large number of conditions and actions. The table header card for the subsequent pages should not be filled in, but, in the case of row continuation (i.e., due to more rules than will fit on a page), a continuation flag—the number "1"—should be set in the RSET rules set entry area of the initial header card. The flag should not be set for multiple pages due to condition and action overflow (i.e., overflow of the entries for a rule or other column onto another page).

The entries in the Table Header are:

Table ID

Each table must be identified by a three-digit number (e.g., 001, 074, 694). This number is nominative only and table ID's need not be sequential nor ordered in any way, but each table ID must be unique within a program. * The Table ID is repeated for every row of a table.

Row No. The Row Number is a three-digit number that for the Table Header is always 000. This designation indicates to the processor that this is a Header Card.

Line. The Line Entry is one alphanumeric character that for the Table Header is always "O". This designation indicates to the processor that this is a Table Header Card.

Rset. The Rules Set (RSET) entry is left blank unless cards are required. If more than one page is used, a numeral "1" is entered in the RSET entry of the Table Header Card. If an entry is made in the RSET entry of the Table Header, a single numeric digit must appear in the RSET entry of all other rows of the decision table.

Table Name. A table may be given any name that conforms to the specifications of COBOL for a procedure-name, usually indicating the function or content of the table. That is, a procedure-name may be composed of alphabetic, numeric, or alphanumeric characters (at least one but not more than thirty) or sets of such characters separated by hyphens (-). However, a hyphen may not begin or end a name. Names must be unique. The Table Name may be referenced by GO TO and

* *Note:* DETAB-X specifications permitted the TABLE ID to be substituted for the Decision Table Name in calls to the table. This has not been implemented in DETAB/65: TABLE ID serves only to distinguish one table from another in sequence checking or EAM processing of cards.

PERFORM operators. The DETAB/65 processor will use this name in generating a COBOL section-name for the table. A dummy paragraph-name must also be generated and inserted after the section-name.

Form. Three basic formats are permitted for decision tables: limited entry (L), extended entry (E), and mixed entry (M). The format of the decision table being specified is indicated by placing one of these three values (L, E, or H) in the Form entry. That is, one alphabetic character.

Cond Rows. A three-digit number containing the number of conditions *rows* (not lines) in the table. That is, this entry specifies the number of conditions that are contained in the table. Leading zeros must be given.

Action Rows. A three-digit number containing the number of action *rows* (not lines) in the table. That is, this entry specifies the number of actions that are contained in the table. Leading zeros must be given.

RULES. A three-digit number specifying the total number of rules in the table. The ELS-rule should be included in the count. Leading zeros must be given.

RULE HEADER

The second header card is the Rule Header used to specify the rule numbers for the table. Besides rule numbers, it will contain the identifying information given for every line of a decision table. As many Rule Header cards will be used as are required to specify all rules.

Table ID. Always the same as the Table Header.

Row No. The Row Number (ROW-NO.) for a Rule Header Card is always 000, indicating that this is a Header card.

Line. The Line (LINE) entry for a Rule Header is always "1," indicating that this is a Rule Header.

RSET

The Rules Set (RSET) entry will be blank except when the number of rules specified requires more than one page. If more than one page is required, the RSET entry will be "1" for the first page (or card) and following pages will be numbered sequentially.

Rule No. A Rule Number must be a three-digit number (e.g., 001, 102, 142, etc.) or ELS. (Every table must have an ELS rule as the last rule to be given.) The first Rule Number is entered in the Rule

Header Card in the first space available beyond the longest entry in the Condition Stub, although blanks may be left between the last character of the longest condition and the first Rule Number. The column that the first digit of the first Rule Number occupies defines the point at which the DETAB/65 processor construes the Condition Entry Area to begin. Spaces occupied by Rule Numbers may vary from three through twelve, but the Rule Number must begin in the leftmost column of the area to be reserved for the rule. The spaces beginning with the first digit of one Rule Number and ending at the last digit before the next is interpreted by the DETAB/65 processor as the number of spaces that are reserved for the condition entries subsumed under that Rule: The number of spaces thus reserved must be equal to or greater than the longest Condition Entry in that rule. The number of spaces reserved may vary from rule to rule, but must not be less than three nor more than twelve. The end of the last rule, and the end of the rules, is indicated by placing a "$" in the first space beyond the last space required by the last rule (i.e., the ELS-rule).* A rule entry may not be split between pages.

If, in an extended entry table, there are not enough columns left on a page to contain all of a particular rule, a new page must be started and an intermediate end-of-rule marker (a "$") set to define the end of the preceding rule and to inform the processor to go on to the next card. When the processor encounters a $, it will check the preceding rule for an ELS-rule. If the preceeding rule is an ELS, search for further rules is stopped and the processor proceeds to the next card. NOTE: Although Rule Numbers should be sequential and entered in ascending order (i.e., 001, 002, 003, etc.), this does not imply that the rules will be executed in that order. The DETAB/65 processor evaluates the matrix of condition entries and optimizes the decision tree for efficient processing.

CONDITION AREA

The cards following the Rule Header Card are used to specify sets of conditions. Each row specifies the states, specific values, or ranges of values that a particular piece of data may assume, or relationships to

* *Note:* An exception is made in the case where the last (ELS) rule includes the 80th column of the card. That is, if RSET = 0, the card column count = 80, the rule header entry is ELS, and the spaces reserved less than or equal to 12, then this is the last rule. Otherwise, an error has occurred.

other data or combinations of these states that the data may assume, and upon which decisions are to be based.

Structurally, a condition consists of two parts: (1) a Condition Stub and (2) a Condition Entry. The Condition Stub area consists of the entries for Table ID, Row No., Line, RSET, and a conditional statement, or portion thereof. The Condition Entry area consists of entries specifying the values of the data or condition specified in the stub that will satisfy the decision requirements of the rules.

The presence of a Condition Area in a table is not required; however, if it is absent, the table can have only one rule in the action area.

TABLE ID

As above, a three-digit entry uniquely identifying a particular decision table.

Row. Row number is a three-digit number used to identify particular conditions, actions, or notes. Condition and action row numbers may vary from 001 to 899; 900 to 999 will designate Notes. Conditions must have lower row numbers than do actions. Row numbers should be, but need not be, assigned in ascending order. Sequential numbering improves legibility of tables for later reference, but the processor may reorder the conditions in sorting to minimize the decision tree. Leading zeros must be filled in, and duplicate row numbers are not permitted.* A row may cover as many lines on a specification form as are required to write out the operators and operands of an expression, up to the maximum permissible value of LINE (i.e., 9).

Line. Line is a one-digit entry used to specify continuations of a row. A Line entry may be blank or range from 1 to 9. Lines, if specified, must be specified in exact sequence because following lines are considered as a continuous description of the condition, action, or comment being given. A blank entry signifies that only one line will be used; if a digit is given, the DETAB/65 processor will look for continuations of the line until the Row Number changes. For limited entry tables, line numbers greater than 1 may be dropped on continuation pages. For extended entry tables, lines not required by the condition entries may be dropped.

* *Note:* Presently, the number of conditions is limited to 50, although row numbers may run much higher.

RSET

The Rules Set entry may be either blank or contain a digit from 1 to 9 or the symbol "$". As specified above, RSET is used to number the sub-tables required to specify all the rules of a table. If more than one sub-table is required, they are numbered sequentially; that is, the RSET entry will be the same for each condition, action, or comment of any sub-table. RSET applies to row continuation horizontally and not vertically. The sheets containing the Condition and Action stubs are considered sub-table 1 no matter how many physical sheets are used. If more rules are required than can be contained in nine sub-tables, a new table must be created, using the same conditions and actions, and to which the ELS-rule may transfer if none of the rules in the first table are satisfied. The $ is used to indicate the end of the table. The entry for the card will contain 9999$.

Deck Sequence. The entries for row, line, and continuation are used in sequencing the deck for presentation to the processor. The expected deck order will be a sort on (1) row, (2) RSET, and (3) Line. The deck make-up is illustrated in Fig. 18-1.

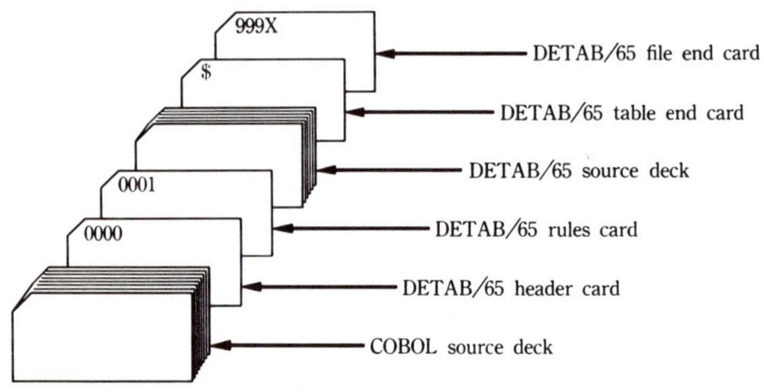

FIG. 18.1 DECK FORMAT FOR DETAB/PROGRAM

Condition Stub. Beginning in column 9 of the specification sheet and continuing for as many lines of a row (up to 9) as are necessary to contain it, any condition that constitutes a legitimate COBOL conditional statement (with the IF implied) may be specified. A condition stub must contan at least one operand. A condition stub entry is bounded by column 9 of the input card format and the first column of the first rule, but may be continued on several lines. A condition stub entry must be contained entirely upon one page (i.e., the first logical subtable) in row-line length, but additional lines may be given on subsequent physical pages. Continuations of lines do not require hyphenation since continuations are treated as part of a single entry. Blanks other than those required because of language specifications are ignored at the beginning and end of lines, permitting the user to indent or organize operands in various ways to increase legibility.

Despite this flexibility in specifying conditions, it is recommended that condition stub entries be kept as short and concise as possible. Any lengthy calculations should be relegated to a separate expression and assigned a name that may be referenced in a condition stub entry.

Entry. A condition entry is specified within the bounds of a rule. That is, it begins in the column containing the leftmost digit of its rule number and ends where the rule immediately to its right begins. A rule must not be less than three nor more than twelve columns in width, but an entry may be continued on the subsequent lines of a row.

Three kinds of condition entries are permitted: Limited, Extended, and Mixed. In a limited entry form, each rule or entry is three columns wide. Permissible entries are (a) Y (i.e., *Yes*), signifying that the stated condition must be true to satisfy the rule; (b) N (i.e., *No*), signifying that the condition must be false to satisfy; (c) - (i.e., a hyphen or blank), signifying that the condition may be either true or false, that the programmer doesn't care which it is. The entry must be made in the second column of the rule and have a blank space both to the right and to the left within the rule.

In extended entry form, leading and following blanks are permissible, but not required. Permissible entries are (a) an operator and an operand, (b) an operand, (c) -, or (d) a blank. In mixed entry format, the entries for any one condition may be either extended or limited, but not both.

In extended and mixed entries, a condition entry too long to be contained within the allotted 12 spaces may be continued on successive lines of a row. Limited entries in a mixed table must still contain a leading and trailing blank and appear in the second column of the rule.

The last rule will be the ELS-rule. The ELS-rule must be given, but all condition entries for the ELS-rule must be blank (or "-").

Neither names nor values may be split between stub and entry. At least one operand or operand and verb must appear in the stub, and at least one condition entry must be given for each condition (except in the case of the "empty" table, in which case no entries are made).

A condition entry may not be split between pages or "sub-tables." (See Rule No., Rule Header section.) If there are less than 12 spaces left on a page and the entry exceeds the available space, the entry must either be made on a new page or sub-table, or multiple lines may be used.

Formula references must not duplicate any data references made in the data division.

On sub-tables beyond the first (i.e., RSET < 1), condition entries must begin in column 9 of the DETAB specification sheet; that is, the first rule number on continuation pages should begin in column 9. Limited entries must still be in the middle column of the entry and extended entries may or may not have leading and trailing blanks.

In all sub-tables, condition entries must start in the first line of a row. For limited entry tables, only Line 1 or a row need be given in sub-tables beyond the first since all condition entries on other lines would be blank. The programmer may wish to retain the vertical alignment, however, to help avoid mistakes. In constructing a condition matrix for a limited entry table all lines except Line 1 will be rejected (i.e., not read into the matrix) and entries so misplaced will be lost.

For extended entries, the programmer may include as many lines as are necessary (up to nine) to express the longest entry; however, this number of lines may vary as required from sub-table to sub-table of the specifications. In writing extended entries, no continuation marks, such as hyphens, need be given since each line is picked up as a continuation of the row. Blanks should therefore be inserted where appropriate to avoid two words being read as one. In setting up the condition matrix, the input editor will continue reading cards into the matrix until a new row number is encountered, creating an image of the input condition matrix area.

In an extended entry table, the programmer may desire to specify a condition such that the value specified is *not* any of the values specified in any other condition entry in that row,. He may do this by specifying "NOT." In the processor this expression will be expanded into an "N" entry and row in combination with every other condition in the row for which a value is given. DON'T CARE entries ("-" or blank) will be ignored. In illustration, consider this example:

Action Area. The action area is used to specify the actions the pro-

FIG. 18-2. EXAMPLE OF THE EXPANSION OF "NOT" AND OF OPTIMIZATION

CONDITION STUB	RULE NO.												
COND.	001	002	003	004	005	006	007	008	009	010	011	012	ELS
001 A = 12	Y	Y	Y	Y	Y	Y	N	N	N	N	N	N	
002 B =	5	10	not	5	10	not	5	10	not	5	10	not	
003 C = OPEN	Y	Y	Y	N	N	N							
004 D = SUNDAY							Y	Y	Y	N	N	N	

This expands to:

	001	002	003	004	005	006	007	008	009	010	011	012	ELS
001 A = 12	Y	Y	Y	Y	Y	Y	N	N	N	N	N	N	
002 B = 5	Y	N	N	Y	N	N	Y	N	N	Y	N	N	
003 B = 10		Y	N		Y	N		Y	N		Y	N	
004 C = OPEN	Y	Y	Y	N	N	N							
005 D = SUNDAY							Y	Y	Y	N	N	N	

After optimization this table will read:

	001	004	002	005	003	006	007	010	008	011	009	012	ELS
001 A = 12	Y	Y	Y	Y	Y	Y	N	N	N	N	N	N	
002 B =	Y	Y	N	N	N	N	Y	Y	N	N	N	N	
003 B = 10			Y	Y	N	N			Y	Y	N	N	
004 C = OPEN	Y	N	Y	N	Y	N							
005 D = SUNDAY							Y	N	Y	N	Y	N	

gram is to take when the conditions satisfying the various rules are met. The actions represent the "THEN" part of the conditional statements for which the conditions are the "IF" portion. All the rules and principles stated for conditions apply equally to the action area. The action area consists of an Action Stub and an Action Entry area just as the Condition Area consisted of a Condition Stub and a Condition Entry Area.

Table ID. The Table ID entry must be the same unique identifier as for the rest of the table.

Row. Row Number entries are a continuation of Row Numbers for the Conditions. However, while the rows need not be numbered in sequential order, it is recommended that they be so ordered, for they will be compiled and executed in the numbered sequence. That is, while condition rows are subject to reordering in the optimization process, actions are not, thus insuring that the program will take a sequence of actions in the order specified by the programmer. That is, the processor will not create any logical errors by reordering actions. (The programmer, however, is free to create his own.)

Line. The Line entry should be left blank if an action can be specified on a single line; otherwise the digits 1, 2, 3, etc., must be used in precise sequence so that the proper actions statement is picked up.

Rset. The RSET entry is left blank if only one sub-table is required. If more than one sub-table is required to contain the additional rules, RSET will be numbered 1, 2, 3, as required, up to 9.

Action Stub. The Action Stub must contain at least an operator. In a limited entry table it will contain the entire action in a single "row" using as many lines as required to specify the entire action to be taken. In extended entry tables an expression may be split between the Action Stub and the Action Entries. Although names and values may be broken and continued from line to line, they may not be split between the Stub and an Entry. This does not prevent subscript values of a name being used as entry values to specify different table destinations resulting from different rules being satisfied.

While no restriction is placed on the kind or length of actions taken (with a procedure or another conditional, if you wish), it is recommended that all lengthy Action Stub and Action Entry entries be defined outside the table and referenced by an appropriate name.

Although, again, no restrictions are imposed, it is recommended that all calls to system (COBOL) subroutines be made in a general fashion. For example, an I/O order should be expressed PERFORM READ,

PERFORM WRITE, etc. This convention will enable the programmer to write a program that is compatible with the manner in which various COBOL I/O modifiers (AT END, EOF, ON SIZE, etc.) have been implemented in different COBOL processors.

Action Entry. For limited entry tables, any action that is associated with a given rule will be indicated by placing an "X" in the central column of the three-column entry space of line 1 (the first line) of the action row. The "X" must be followed and preceded by a blank.

Extended action entries will be treated in the same manner as extended condition entries. Care will be taken in generating COBOL statements not to disturb the basic operating sequence specified by the programmer.

NOTES

A note is a descriptive statement that has no functional significance and that carries the same uses and restrictions as notes in COBOL-61. A note has the same Table ID as the table that it is applied to. Notes are given a row number of 900 to 999 and note lines are numbered 1, 2, 3, etc., in order for other rows and lines to maintain their sequential order. Notes may be written anywhere in the table except between the Table Header and the Rule Header Cards. When the processor encounters a note row designator, it will transfer the card into the COBOL code area as a note without further processing.

Note, however, that as presently constituted, all notes would be sorted to the end of the table by preprocessing to put the cards in order. If it is desired that notes be inserted in place, the cards may be hand-ordered if more than one page or sub-table is used, or not sorted if only one sub-table (the first) is used.

RESTRICTIONS AND USAGES

Certain restrictions and preculiarities of usage become apparent as decision table specifications are used in combination with the COBOL languages. These restrictions are, in general, not serious, but minor logical errors may be avoided if the programmer is aware of them.

LINKAGES

A decision table may be entered in a variety of ways. The preferred mode of entry is through a PERFORM verb, in which case a linkage back to the main control sequence may be generated. If the table is to be used iteratively on a succession of inputs or cases (as in a scan routine), then PERFORM is the logical choice. If tables are nested, the PERFORM verb should be used to step from one table to the next and back. If a table may be entered from any one of a set of points, the PERFORM verb and RETURN entry are most convenient.

A table may be entered through a GO TO, but in this case the processor has no means of knowing what location to return to in the main control sequence unless this location is explictly stated in conjunction with the separate rules. Of course, if one of the functions of the table is to switch control to various program regions depending upon the conditions encountered, this may be the preferred mode of specification. A sequence of tables may be stepped through with a series of GO TO's, but it is felt that this mode is inferior to the use of PERFORM's.

Note that a table may not be entered via an ENTER verb, nor may an ENTER verb be used within a table without creating difficulties. All information necessary for the processing of the table and its COBOL statements is contained in the header cards.

Note also that tables cannot be entered at any point except at the beginning. If entry is desired at intermediate points in a set of conditions, this effect may be achieved by creating a sequence of tables and chaining them together. If tables are nested, however, returns to points in the Action Area of previous tables may be made through the operation of the PERFORM verb and normal exits, or may be specified directly by a GO TO.

Note that if a GO TO table-name-1 is given within the range of a COBOL PERFORM . . . THRU that the sequence of control may be lost unless the programmer has established instructions that will get the sequence back within the loop, However, no such trouble should be encountered when a PERFORM table-name-1 is given unless the table contains only explicit GO TO's that pick up the sequence elsewhere.

```
1  0000 TABLEXX                         L 004001003
   0001    001002ELS$
        C1  N  N
        C2     N
        C3     N
        C4  Y  N
        A1  X
      $
1  0000 TABLEXXX                        L 004001004
   0001    001002003ELS$
        C1  Y  Y  Y
        C2     Y  N
        C3     N  N
        C4  Y  N  N
        A1  X
      $
1  0000 TABLEXXXX                       L 006001004
   0001    001002003ELS$
        C1  Y  Y  Y
        C2  Y     N
        C3  N  N  N
        C4  N     N
        C5  N     N
        C6  N  Y  N
        A1  X  X  X
      $
1  0000   TEST-001                      L 003001004
   0001            001002003ELS$
            C-1     Y  Y  N
            C-2     Y  Y  Y
            C-3     Y  N  Y
          ACTION-1  X  X  X  X
      $
1  0000   TEST-002                      L 002001005
   0001            001002003004ELS$
            C-1     Y  N  Y  N
            C-2     N  Y  Y  N
          ACTION-1  X  X  X  X  X
      $
1  0000   TEST-003                      L 003001009
   0001            001002003004005006007008ELS$
            C-1     Y  Y  Y  N  Y  N  N  N
            C-2     Y  Y  N  Y  N  N  Y  N
            C-3     Y  N  Y  Y  N  Y  N  N
          ACTION-1  X  X  X  X  X  X  X  X  X
      $
1  0000   TEST-004                      L 004001017
   0001            001002003004005006007008009010011012013014015016ELS$
            C-1     Y  Y  Y  Y  N  Y  Y  N  Y  N  N  Y  N  N  N  N
            C-2     Y  Y  Y  N  Y  Y  N  N  N  Y  Y  N  N  N  Y  N
            C-3     Y  Y  N  Y  Y  N  N  Y  Y  N  Y  N  N  Y  N  N
            C-4     Y  N  Y  Y  Y  N  Y  Y  N  Y  N  N  Y  N  N  N
          ACTION-1  X  X  X  X  X  X  X  X  X  X  X  X  X  X  X  X  X
      $
1  0000   TEST-005                      L 004001003
   0001            001002ELS$
            C-1     Y  Y
            C-2     Y  Y
            C-3     Y  Y
            C-4     Y  N
          ACTION-1  X  X  X
```

```
          $
1       0000    TEST-006                              L 004001004
        0001                 001002003ELS$
                  C-1         Y  Y  Y
                  C-2         Y  Y  Y
                  C-3         Y  Y  N
                  C-4         Y  N  Y
                ACTION-1      X  X  X  X
          $
1       0000    TEST-007                              L 004001005
        0001                 001002003004ELS$
                  C-1         Y  Y  N  -
                  C-2         Y  Y  Y  Y
                  C-3         Y  Y  Y  N
                  C-4         Y  N  Y  Y
                ACTION-1      X  X  X  X  X
          $
1       0000    TEST-009                              L 003001009
        0001                 001002003004005006007008ELS$
                  C-1         Y  Y  N  N  Y  N  N  Y
                  C-2         Y  Y  Y  N  N  N  Y  N
                  C-3         Y  N  Y  Y  Y  N  N  N
                ACTION-1      X  X  X  X  X  X  X  X  X  X
          $
1       0000    TEST-010                              L 007001009
        0001                 001002003004005006007008ELS$
                  C-1         Y  Y  Y  Y  N  N  N  N
                  C-2         Y  Y  N  N  Y  N  Y  N
                  C-3         Y  N  Y  N  N  Y  Y  N
                ACTION-1      X  X  X  X  X  X  X  X  X
          $
1       0000    TEST-011                              L 005001005
        0001                 001002003003ELS$
                  C-1         Y  Y  Y  N
                  C-2         Y  Y  Y  Y
                  C-3         Y  Y  Y  Y
                  C-4         Y  Y  N  Y
                  C-5         Y  N  Y  Y
                ACTION-1      X  X  X  X  X
          $
1       0000    TEST-012                              L 003001009
        0001                 001002003004005006007008ELS$
                  C-1         N  N  N  N  Y  Y  Y  Y
                  C-2         N  N  Y  N  Y  N  N  Y
                  C-3         N  Y  N  Y  N  Y  N  Y
                ACTION-1      X  X  X  X  X  X  X  X  X
          $
1       0000    TEST-013                              L 005001005
        0001                 001002003004ELS$
                  C-1         Y  Y  Y  Y
                  C-2         Y  Y  Y  N
                  C-3         Y  Y  Y  Y
                  C-4         Y  Y  N  Y
                  C-5         Y  N  Y  Y
                ACTION-1      X  X  X  X  X
          $
1       0000    TEST-014                              L 010001010
        0001                 001002003004005006007008009ELS$
                  C-1         Y  N  Y  Y  N  Y  Y  N  N
                  C-2         Y  Y  Y  Y  Y  -  Y  -  N
                  C-3         Y  Y  Y  Y  Y  N  Y  N  Y
                  C-4         Y  Y  Y  Y  Y  N  Y  N  Y
                  C-5         Y  Y  Y  Y  Y  Y  Y  Y  Y
                  C-6         Y  Y  -  Y  N  -  N  N  -
                  C-7         Y  Y  -  Y  N  -  -  N  -
```

```
                       C-8      Y   Y   N   Y   Y   Y   Y   N   Y
                       C-9      Y   Y   Y   Y   Y   Y   Y   Y   Y
                       C-10     Y   Y   Y   N   Y   Y   Y   Y   Y
                     ACTION-1   X   X   X   X   X   X   X   X   X   X
            $
 1      0000         TEST-015                               L 010001010
        0001                  001002003004005006007008009ELS$
                       C-1      Y   Y   Y   N   Y   Y   Y   Y   Y
                       C-2      Y   Y   Y   Y   Y   Y   Y   Y   Y
                       C-3      Y   Y   N   Y   Y   Y   Y  'N   Y
                       C-4      Y   Y   -   Y   N   -   -   N   -
                       C-5      Y   Y   -   Y   N   -   N   N   -
                       C-6      Y   Y   Y   Y   Y   Y   Y   Y   Y
                       C-7      Y   Y   Y   Y   Y   N   Y   N   Y
                       C-8      Y   Y   Y   Y   Y   N   Y   N   Y
                       C-9      Y   Y   Y   Y   Y   -   Y   -   N
                       C-10     Y   N   Y   Y   N   Y   Y   N   N
                     ACTION-1   X   X   X   X   X   X   X   X   X   X
            $
 1      0000         TEST-016                               L 005001005
        0001                  001002003004ELS$
                       C-1      Y   Y   Y   Y
                       C-2      N   Y   Y   Y
                       C-3      -   N   Y   Y
                       C-4      -   -   N   Y
                       C-5      -   -   -   N
                     ACTION-1   X   X   X   X   X
            $
 1      0000         TEST-017                               L 005001005
        0001                  001002003004ELS$
                       C-1      Y   Y   Y   Y
                       C-2      Y   Y   Y   N
                       C-3      Y   Y   N   -
                       C-4      Y   N   -   -
                       C-5      N   -   -   -
                     ACTION-1   X   X   X   X   X
            $
 1      0000         TEST-018 (CHOISE-PICK)          L 010011010
        0001                                   001002003004005006007008009ELS$
                     C1                          Y   Y   Y   N   Y   Y   Y   Y   Y
                     C2                          Y   Y   Y   Y   Y   Y   Y   Y   Y
                     C3                          Y   Y   N   Y   Y   Y   Y   N   Y
                     C4                          Y   Y       Y   N           N
                     C5                          Y   Y       Y   N           N
                     C6                          Y   Y   Y   Y   Y   Y   Y   Y   Y
                     C7                          Y   Y   Y   Y   Y   N   Y   N   Y
                     C8                          Y   Y   Y   Y   Y   N   Y   N   Y
                     C9                          Y   Y   Y   Y   Y       Y       N
                     C10                         Y   N   Y   Y   N   Y   Y   N   N
                     A1                          X   X   X   X   X   X   X   X   X   X
                     A2                                  X           X
                     A3                                  X           X
                     A4                                  X           X
                     A5                                          X               X
                     A6                                          X               X
                     A7                                          X               X
                     A8                          X   X                   X   X
                     A9                          X   X                   X   X
                     A10                         X   X                   X   X
                     A11                                     X                       X
            $
        999X
```

```
IDENTIFICATION DIVISION.                                          DETAB-65
PROGRAM-ID. PREPROCESSOR FOR DETAB-65.                            DETAB-65
AUTHOR. ANSON CHAPMAN.                                            DETAB-65
DATE-WRITTEN. 12/30/64.                                           DETAB-65
DATE-COMPILED.                                                    DETAB-65
REMARKS.                                                          DETAB-65
    THE GENERATOR PORTION OF THE PREPROCESSOR ANALIZES A          DETAB-65
    DECISION TABLE AND GENERATES SIMPLE CONDITIONAL STATEMENTS    DETAB-65
    FOR Y'S, N'S AND BLANKS AND WILL GENERATE IF STATEMENTS FOR   DETAB-65
    ONE PATH THRU THE TREE THE ACTION CORRESPONDING TO THE PATH   DETAB-65
    IS GENERATED IN STMTS DX014 THRU DX032 THIS PATH IS DELETED   DETAB-65
    FROM THE TREE IN DX016 THRU DX020 DX301 THRU DX061            DETAB-65
    REINITIALIZES THE TREE, FINDS THE LAST NODE CONNECTED TO      DETAB-65
    THIS PATH AND COMES BACK TO DX003 FOR ANOTHER PASS THRU THE   DETAB-65
    NEXT PATH THIS PROCESS IS REPEATED UNTIL IF STATEMENTS        DETAB-65
    HAVE BEEN GENERATED FOR ALL PATHS THRU THE DECISION TABLE     DETAB-65
    TREE STRUCTURE.                                               DETAB-65
ENVIRONMENT DIVISION.                                             DETAB-65
CONFIGURATION SECTION.                                            DETAB-65
SOURCE-COMPUTER.  CONTROL DATA 1604A.                             DETAB-65
OBJECT-COMPUTER.  CONTROL DATA 1604A.                             DETAB-65
SPECIAL-NAMES.                                                    DETAB-65
    SYSTEM-INPUT-TAPE IS SIT.                                     DETAB-65
INPUT-OUTPUT SECTION.                                             DETAB-65
FILE-CONTROL.                                                     DETAB-65
    SELECT CARD-INPUT, ASSIGN TO SYSTEM-INPUT-TAPE MULTIPLE REEL. DETAB-65
    SELECT CARD-OUTPUT, ASSIGN TO SYSTEM-PUNCH-TAPE.              DETAB-65
    SELECT LIST-OUTPUT,.ASSIGN TO SYSTEM-OUTPUT-TAPE.             DETAB-65
DATA DIVISION.                                                    DETAB-65
FILE SECTION.                                                     DETAB-65
FD CARD-INPUT                                                     DETAB-65
    LABEL RECORDS ARE OMITTED,                                    DETAB-65
    DATA RECORDS ARE TEST-CARD.                                   DETAB-65
01 TEST-CARD.                                                     DETAB-65
    02 FILLER  PICTURE X(80).                                     DETAB-65
FD CARD-OUTPUT                                                    DETAB-65
    LABEL RECORDS ARE OMITTED,                                    DETAB-65
    DATA RECORDS ARE CRD-OUT, DETAB-CRD, DUM-1.                   DETAB-65
01 CRD-OUT.                                                       DETAB-65
    02 FILLER  PICTURE X(7).                                      DETAB-65
    02 BODY.                                                      DETAB-65
      03 FILLER  PICTURE X(4).                                    DETAB-65
      03 B-MARG  PICTURE X(61).                                   DETAB-65
    02 IDFLD   PICTURE X(8).                                      DETAB-65
01 DETAB-CRD.                                                     DETAB-65
    02 FILLER  PICTURE XXX.                                       DETAB-65
    02 IDENT.                                                     DETAB-65
      03 ROW-NO  PICTURE 999.                                     DETAB-65
      03 LINE-ID PICTURE X.                                       DETAB-65
    02 FILLER  PICTURE X(73).                                     DETAB-65
01 DUM-1.                                                         DETAB-65
    02 CRD-COL PICTURE X      OCCURS 80 TIMES.                    DETAB-65
FD LIST-OUTPUT                                                    DETAB-65
    LABEL RECORDS ARE OMITTED,                                    DETAB-65
    DATA RECORD IS TAPE-LIST.                                     DETAB-65
01 TAPE-LIST.                                                     DETAB-65
    02 FILLER  PICTURE X(11).                                     DETAB-65
    02 CARDX   PICTURE 999.                                       DETAB-65
    02 FILLER  PICTURE X(66).                                     DETAB-65
WORKING-STORAGE SECTION.                                          DETAB-65
77 AZ          PICTURE XX     VALUE 'AZ'.                         DETAB-65
77 CARDCNT     PICTURE 999    COMPUTATIONAL SYNCHRONIZED RIGHT.   DETAB-65
```

```
77 COLIX      PICTURE 999    COMPUTATIONAL SYNCHRONIZED RIGHT.   DETAB-65
77 COLUM      PICTURE 999    COMPUTATIONAL SYNCHRONIZED RIGHT.   DETAB-65
77 DUMIX      PICTURE 999    COMPUTATIONAL SYNCHRONIZED RIGHT.   DETAB-65
77 ELMCT      PICTURE 999    COMPUTATIONAL SYNCHRONIZED RIGHT.   DETAB-65
77 ELMCX      PICTURE 999    COMPUTATIONAL SYNCHRONIZED RIGHT.   DETAB-65
77 ELMRX      PICTURE 999    COMPUTATIONAL SYNCHRONIZED RIGHT.   DETAB-65
77 EXIX       PICTURE 999    COMPUTATIONAL SYNCHRONIZED RIGHT.   DETAB-65
77 KEY-1      PICTURE 999    COMPUTATIONAL SYNCHRONIZED RIGHT.   DETAB-65
77 KEY-2      PICTURE 999    COMPUTATIONAL SYNCHRONIZED RIGHT.   DETAB-65
77 KEY-3      PICTURE 999    COMPUTATIONAL SYNCHRONIZED RIGHT.   DETAB-65
77 LABIX      PICTURE 999    COMPUTATIONAL SYNCHRONIZED RIGHT.   DETAB-65
77 LABNO      PICTURE 999    COMPUTATIONAL SYNCHRONIZED RIGHT.   DETAB-65
77 NACTS      PICTURE 999    COMPUTATIONAL SYNCHRONIZED RIGHT.   DETAB-65
77 NCOLS      PICTURE 999    COMPUTATIONAL SYNCHRONIZED RIGHT.   DETAB-65
77 NORLS      PICTURE 999    COMPUTATIONAL SYNCHRONIZED RIGHT.   DETAB-65
77 NOCON      PICTURE 999    COMPUTATIONAL SYNCHRONIZED RIGHT.   DETAB-65
77 NRLS       PICTURE 999    COMPUTATIONAL SYNCHRONIZED RIGHT.   DETAB-65
77 NROWS      PICTURE 999    COMPUTATIONAL SYNCHRONIZED RIGHT.   DETAB-65
77 ROWIX      PICTURE 999    COMPUTATIONAL SYNCHRONIZED RIGHT.   DETAB-65
01 DUM-2.                                                        DETAB-65
    02 FILLER          OCCURS 50 TIMES.                          DETAB-65
       03 STRTCOL PICTURE 99  COMPUTATIONAL SYNCHRONIZED RIGHT.  DETAB-65
       03 NMCOLS  PICTURE 99  COMPUTATIONAL SYNCHRONIZED RIGHT.  DETAB-65
01 DUM-3.                                                        DETAB-65
    02 COLS       PICTURE X      OCCURS 12 TIMES.                DETAB-65
01 DUM-4.                                                        DETAB-65
    02 EGOTO      PICTURE X      OCCURS 5 TIMES.                 DETAB-65
01 DUM-5.                                                        DETAB-65
    02 TEMP       PICTURE X      OCCURS 58 TIMES.                DETAB-65
01 DUM-10  PICTURE X(8)  VALUE 'SECTION.'.                       DETAB-65
01 DUM-12  REDEFINES DUM-10.                                     DETAB-65
    02 NMSEC      PICTURE X      OCCURS 8 TIMES.                 DETAB-65
01 HEADER.                                                       DETAB-65
    02 FILLER     PICTURE X(8).                                  DETAB-65
    02 TBLNME     PICTURE X(30).                                 DETAB-65
    02 FORMID     PICTURE XX.                                    DETAB-65
    02 NCOND      PICTURE 9(3).                                  DETAB-65
    02 ACTNS      PICTURE 9(3).                                  DETAB-65
    02 NORULS     PICTURE 9(3).                                  DETAB-65
    02 FILLER     PICTURE X(51).                                 DETAB-65
01 DPRINT.                                                       DETAB-65
    02 DLABEL.                                                   DETAB-65
       03 FILLER  PICTURE X(7)      VALUE SPACES.                DETAB-65
       03 DUM-6.                                                 DETAB-65
          04 LABNM PICTURE XX.                                   DETAB-65
          04 LABVL PICTURE 9(3).                                 DETAB-65
       03 FILLER  PICTURE X         VALUE '.'.                   DETAB-65
    02 DGOTO.                                                    DETAB-65
       03 FILLER  PICTURE A(7)      VALUE ' GO TO '.             DETAB-65
       03 DGOLN.                                                 DETAB-65
          04 DGOLB PICTURE XX.                                   DETAB-65
          04 DGONO PICTURE 999.                                  DETAB-65
    02 HOUSTON.                                                  DETAB-65
       03 CNDI    PICTURE X(58) OCCURS 50 TIMES.                 DETAB-65
       03 ATBL    PICTURE X(58) OCCURS 50 TIMES.                 DETAB-65
01 LINE1.                                                        DETAB-65
    02 FILLER     PICTURE X(14)     VALUE               IF       DETAB-65
    02 COND       PICTURE X(58).                                 DETAB-65
01 TEXAS.                                                        DETAB-65
    02 LINE2.                                                    DETAB-65
       03 FILLER  PICTURE A(11).                                 DETAB-65
       03 CDOPR   PICTURE X(12).                                 DETAB-65
       03 PIF     PICTURE X.                                     DETAB-65
       03 DELSE   PICTURE X(6).                                  DETAB-65
       03 ELOPR   PICTURE X(12).                                 DETAB-65
       03 PELSE   PICTURE X.                                     DETAB-65
```

```
        03 FILLER PICTURE A(29).                                        DETAB-65
      02 LINE3 REDEFINES LINE2.                                         DETAB-65
        03 FILLER PICTURE X(7).                                         DETAB-65
        03 DNAME.                                                       DETAB-65
          04 TCOLS PICTURE X OCCURS 58 TIMES                            DETAB-65
        03 FILLER PICTURE X(7).                                         DETAB-65
      02 FILLER REDEFINES LINE2.                                        DETAB-65
        03 FILLER PICTURE X(11).                                        DETAB-65
        03 BNAME PICTURE X(58).                                         DETAB-65
        03 FILLER PICTURE XXX.                                          DETAB-65
      02 DECISION-TABLE.                                                DETAB-65
        03 ROW                 OCCURS 50 TIMES.                         DETAB-65
           04 COLMN PICTURE X    OCCURS 100 TIMES.                      DETAB-65
 01 ELIMT.                                                              DETAB-65
      02 ELIMC PICTURE 999   OCCURS 25 TIMES.                           DETAB-65
 01 MATIT.                                                              DETAB-65
      02 MATIX PICTURE 999   OCCURS 25 TIMES.                           DETAB-65
 01 MICDESCR.                                                           DETAB-65
      02 POPUL PICTURE 999   OCCURS 128 TIMES.                          DETAB-65
      02 SAVCL PICTURE X       OCCURS 25 TIMES.                         DETAB-65
 01 WRNING-PRINT.                                                       DETAB-65
      02 FILLER PICTURE X(17)  VALUE                                    DETAB-65
          ' ****** WARNING. '                                           DETAB-65
      02 WRNING-IMAGE PICTURE X(52).                                    DETAB-65
 01 WARNING-MESSAGES.                                                   DETAB-65
      02 WRNING-1 PICTURE X(52)  VALUE                                  DETAB-65
          'NO ELSE RULE CARD. LAST RULE PROCESSED AS ELSE RULE.'.       DETAB-65
      02 WRNING-2 PICTURE X(31)  VALUE                                  DETAB-65
          'REDUNDANCY. CHECK THESE RULES                                DETAB-65
 01 ERR-PRNT.                                                           DETAB-65
      02 FILLER    PICTURE X(30)  VALUE                                 DETAB-65
           ****** ERROR. TABLE SKIPPED                                  DETAB-65
      02 ERR-IMAGE PICTURE X(53).                                       DETAB-65
 01 ERROR-MESSAGES.                                                     DETAB-65
      02 ERR-1       PICTURE X(48). VALUE                               DETAB-65
          'PRESENTLY, TABLES RESTRICTED TO LIMITED ENTRIES.'.           DETAB-65
      02 ERR-2       PICTURE X(42)  VALUE                               DETAB-65
          'TABLE-NAME MISSING FROM TABLE HEADER CARD.'.                 DETAB-65
      02 ERR-3       PICTURE X(19)  VALUE                               DETAB-65
          'RULES CARD MISSING. .                                        DETAB-65
      02 ERR-4       PICTURE X(39)  VALUE                               DETAB-65
          'LESS THAN THREE RULE COLUMNS SPECIFIED.'.                    DETAB-65
      02 ERR-5       PICTURE X(43)  VALUE                               DETAB-65
          'PRESENTLY, CONTINUED RULES NOT IMPLEMENTED.'.                DETAB-65
      02 ERR-6       PICTURE X(40)  VALUE                               DETAB-65
          'CONDITION STUB ENTRY EXCEEDS 58 COLUMNS.'.                   DETAB-65
      02 ERR-7       PICTURE X(26)  VALUE                               DETAB-65
          'MORE THAN 12 RULE COLUMNS. .                                 DETAB-65
      02 ERR-8       PICTURE X(53)  VALUE                               DETAB-65
          'NUMBER OF RULES ENCOUNTERED DISAGREES WITH RULE CARD.'.      DETAB-65
      02 ERR-9     PICTURE X(41)   VALUE                                DETAB-65
          'MORE THAN 50 ACTION OR CONDITION ENTRIES.'.                  DETAB-65
      02 ERR-10    PICTURE X(46)   VALUE                                DETAB-65
          'DECISION TABLE LOGIC ERROR. PROCESSING HALTED.'.             DETAB-65
 PROCEDURE DIVISION.                                                    DETAB-65
 DETAB65.                                                               DETAB-65
      OPEN INPUT CARD-INPUT, OUTPUT CARD-OUTPUT, LIST-OUTPUT.           DETAB-65
 DT001.                                                                 DETAB-65
      PERFORM READ-1.                                                   DETAB-65
      IF '0000' = IDENT OF DETAB-CRD GO TO MONITOR.                     DETAB-65
      WRITE DETAB-CRD.                                                  DETAB-65
      GO TO DT001.                                                      DETAB-65
 MONITOR.                                                               DETAB-65
      MOVE DETAB-CRD TO HEADER.                                         DETAB-65
      IF TBLNME = SPACES GO TO EM02.                                    DETAB-65
      IF FORMID OF HEADER NOT = 'L' GO TO EM01.                         DETAB-65
```

```
     MOVE SPACES TO HOUSTON, TEXAS.                                    DETAB-65
     MOVE ZEROES TO DUM-2.                                             DETAB-65
     MOVE TBLNME TO DUM-5, DNAME.                                      DETAB-65
     PERFORM RSCAN.                                                    DETAB-65
     PERFORM DT005 VARYING EXIX FROM 1 BY 1 UNTIL EXIX = 9.            DETAB-65
     PERFORM READ-1.                                                   DETAB-65
     IF IDENT OF DETAB-CRD NOT = '0001' GO TO EM03.                    DETAB-65
                                                                       DETAB-65
     NOTE RULES CONVERSION SECTION.                                    DETAB-65
                                                                       DETAB-65
     MOVE 0 TO CARDCNT.                                                DETAB-65
     MOVE 1 TO NRLS.                                                   DETAB-65
     MOVE 9 TO COLUM, STRTCOL (NRLS).                                  DETAB-65
 DT050.                                                                DETAB-65
     IF CRD-COL (COLUM) = SPACE GO TO DT053.                           DETAB-65
     IF CARDCNT IS LESS THAN 3 GO TO EM04.                             DETAB-65
     MOVE CARDCNT TO NMCOLS (NRLS).                                    DETAB-65
     IF CRD-COL (COLUM) = '$' GO TO DT055.                             DETAB-65
     ADD 1 TO NRLS.                                                    DETAB-65
     MOVE COLUM TO STRTCOL (NRLS).                                     DETAB-65
     MOVE 3 TO CARDCNT.                                                DETAB-65
     ADD 3 TO COLUM.                                                   DETAB-65
     IF COLUM IS GREATER THAN 80 GO TO EM05.                           DETAB-65
     GO TO DT050.                                                      DETAB-65
 DT005.                                                                DETAB-65
     MOVE NMSEC (EXIX) TO TCOLS (DUMIX).                               DETAB-65
     ADD 1 TO DUMIX.                                                   DETAB-65
 DT053.                                                                DETAB-65
     ADD 1 TO CARDCNT, ADD 1 TO COLUM.                                 DETAB-65
     IF CARDCNT IS NOT GREATER THAN 12 GO TO DT050.                    DETAB-65
     IF CARDCNT IS GREATER THAN 58 GO TO EM06.                         DETAB-65
     IF NRLS = 1 GO TO DT050 ELSE GO TO EM07.                          DETAB-65
 DT055.                                                                DETAB-65
     SUBTRACT 1 FROM NMCOLS (NRLS), SUBTRACT 1 FROM NRLS.              DETAB-65
     IF NRLS NOT = NORJLS GO TO EM08.                                  DETAB-65
                                                                       DETAB-65
     NOTE DETAB CARD SECTION.                                          DETAB-65
                                                                       DETAB-65
     ADD 1 TO NRLS.                                                    DETAB-65
     MOVE STRTCOL (NRLS) TO COLUM.                                     DETAB-65
     IF CRD-COL (COLUM) = 'E' GO TO DT056.                             DETAB-65
     MOVE WRNING-1 TO WRNING-IMAGE.                                    DETAB-65
     WRITE TAPE-LIST FROM WRNING-PRINT.                                DETAB-65
 DT056.                                                                DETAB-65
     MOVE 1 TO KEY-2, KEY-3, ROWIX.                                    DETAB-65
 DT057.                                                                DETAB-65
     PERFORM READ-1.                                                   DETAB-65
     IF ROW-NO OF DETAB-CRD =  999  GO TO DT057.                       DETAB-65
     MOVE 1 TO KEY-1, COLIX.                                           DETAB-65
     IF LINE-ID OF DETAB-CRD = '$' GO TO TBLPROC.                      DETAB-65
     MOVE STRTCOL (KEY-1) TO COLUM.                                    DETAB-65
                                                                       DETAB-65
     NOTE CONDACT SECTION.                                             DETAB-65
                                                                       DETAB-65
     MOVE SPACES TO DUM-5.                                             DETAB-65
     MOVE 1 TO EXIX.                                                   DETAB-65
 CONDACT.                                                              DETAB-65
     MOVE CRD-COL (COLUM) TO TEMP (EXIX).                              DETAB-65
     IF EXIX GREATER NMCOLS (KEY-1) GO TO DT057-1.                     DETAB-65
     ADD 1 TO EXIX, ADD 1 TO COLUM, GO TO CONDACT.                     DETAB-65
 DT057-1.                                                              DETAB-65
     IF KEY-2 IS GREATER THAN 50 GO TO EM09.                           DETAB-65
     IF KEY-2 IS GREATER THAN NCOND GO TO DT058.                       DETAB-65
     MOVE DUM-5 TO CNDI (KEY-2).                                       DETAB-65
     ADD 1 TO KEY-2.                                                   DETAB-65
```

```
    GO TO DT059.                                                  DETAB-65
DT058.                                                            DETAB-65
    IF KEY-3 IS GREATER THAN 50 GO TO EM09.                       DETAB-65
    MOVE DUM-5 TO ATBL (KEY-3).                                   DETAB-65
    ADD 1 TO KEY-3.                                               DETAB-65
DT059.                                                            DETAB-65
    PERFORM DT060 THRU DT061 VARYING KEY-1 FROM 2 BY 1 UNTIL      DETAB-65
      KEY-1 IS GREATER THAN NRLS.                                 DETAB-65
    ADD 1 to ROWIX.                                               DETAB-65
    GO TO DT057.                                                  DETAB-65
DT060.                                                            DETAB-65
    MOVE STRTCOL (KEY-1) TO COLUM.                                DETAB-65
                                                                  DETAB-65
    NOTE VARAMOVE SECTION.                                        DETAB-65
                                                                  DETAB-65
    MOVE SPACES TO DUM-3.                                         DETAB-65
    MOVE 1 TO EXIX.                                               DETAB-65
VARAMVE.                                                          DETAB-65
    MOVE CRD-COL (COLUM) TO COLS (EXIX).                          DETAB-65
    IF EXIX GREATER NMCOLS (KEY-1) GO TO DT060-                   DETAB-65
    ADD 1 TO EXIX, ADD 1 TO COLUM, GO TO VARAMVE.                 DETAB-65
DT060-1.                                                          DETAB-65
    EXAMINE DUM-3 REPLACING ALL '-' BY SPACES.                    DETAB-65
    IF DUM-3 = SPACES GO TO DT061.                                DETAB-65
    EXAMINE DUM-3 TALLYING UNTIL FIRST 'N'.                       DETAB-65
    IF TALLY = 12 MOVE 'Y' TO COLMN (ROWIX, COLIX) ELSE           DETAB-65
      MOVE 'N' TO COLMN (ROWIX, COLIX).                           DETAB-65
DT061.                                                            DETAB-65
    ADD 1 TO COLIX.                                               DETAB-65
TBLPROC.                                                          DETAB-65
    PERFORM L2OUT THRU RITAB.                                     DETAB-65
    MOVE 'DX000' TO DUM-6.                                        DETAB-65
    PERFORM DLOUT THRU RITAB.                                     DETAB-65
                                                                  DETAB-65
    NOTE DECISION SECTION.                                        DETAB-65
                                                                  DETAB-65
    MOVE ZERO TO LABIX, LABNO.                                    DETAB-65
    MOVE ACTNS TO NACTS.                                          DETAB-65
    COMPUTE NORLS = NORULS - 1.                                   DETAB-65
    MOVE NCOND TO NOCON.                                          DETAB-65
    PERFORM DX042 VARYING COLIX FROM 1 BY 1 UNTIL COLIX = NORLS.  DETAB-65
DX042.                                                            DETAB-65
    MOVE COLIX TO MATIX (COLIX).                                  DETAB-65
DX001.                                                            DETAB-65
    PERFORM DX002 VARYING COLIX FROM 1 BY 1 UNTIL COLIX = NORLS.  DETAB-65
DX002.                                                            DETAB-65
    MOVE COLIX TO ELIMC (COLIX).                                  DETAB-65
DX050.                                                            DETAB-65
    MOVE NOCON TO NROWS.                                          DETAB-65
    MOVE NORLS TO NCOLS.                                          DETAB-65
    MOVE 0 TO ROWIX.                                              DETAB-65
    GO TO DX004.                                                  DETAB-65
DX003.                                                            DETAB-65
    PERFORM L1OUT THRU RITAB.                                     DETAB-65
    PERFORM L2OUT THRU RITAB.                                     DETAB-65
DX004.                                                            DETAB-65
    MOVE SPACES TO LINE2.                                         DETAB-65
DX005.                                                            DETAB-65
    ADD 1 TO ROWIX.                                               DETAB-65
    MOVE ZERO TO DUMIX.                                           DETAB-65
    IF ROWIX = NOCON GO TO DX014.                                 DETAB-65
    MOVE 1 TO COLIX.                                              DETAB-65
                                                                  DETAB-65
    NOTE  ARE THERE ALL BLANKS IN THIS ROW.                       DETAB-65
                                                                  DETAB-65
DX005-1.                                                          DETAB-65
```

```
     IF COLIX GREATER NCOLS GO TO DX005-2.                            DETAB-65
     MOVE ELIMC (COLIX) TO ELMCX.                                     DETAB-65
     IF COLMN (ROWIX, ELMCX) = ' ' OR 'B'                             DETAB-65
         NEXT SENTENCE ELSE GO TO DX051.                              DETAB-65
     ADD 1 TO COLIX.                                                  DETAB-65
     GO TO DX005-1.                                                   DETAB-65
 DX005-2.                                                             DETAB-65
     PERFORM DX400 THRU DX402 VARYING COLIX FROM 1 BY 1               DETAB-65
       UNTIL COLIX IS GREATER THAN NCOLS.                             DETAB-65
     GO TO DX005.                                                     DETAB-65
 DX400.                                                               DETAB-65
     MOVE ELIMC (COLIX) TO ELMCT.                                     DETAB-65
     MOVE 1 TO ELMRX.                                                 DETAB-65
 DX400-1.                                                             DETAB-65
     IF ELMRX = ROWIX GO TO DX400-2.                                  DETAB-65
     IF COLMN (ELMRX, ELMCT) = ' '                                    DETAB-65
        MOVE 'B' TO COLMN (ROWIX, ELMCT)                              DETAB-65
        GO TO DX402.                                                  DETAB-65
     ADD 1 TO ELMRX.                                                  DETAB-65
     GO TO DX400-1.                                                   DETAB-65
 DX400-2.                                                             DETAB-65
     MOVE 'Y' TO COLMN (ROWIX, ELMCT).                                DETAB-65
 DX402.                                                               DETAB-65
     EXIT.                                                            DETAB-65
 DX051.                                                               DETAB-65
     MOVE CNDI (ROWIX) TO COND.                                       DETAB-65
                                                                      DETAB-65
     NOTE  IS THERE A Y OR N IN THIS ROW.                             DETAB-65
                                                                      DETAB-65
     MOVE 1 TO COLIX.                                                 DETAB-65
 DX051-1.                                                             DETAB-65
     IF COLIX GREATER NCOLS GO TO DX051-2.                            DETAB-65
     MOVE ELIMC (COLIX) TO ELMCX.                                     DETAB-65
     IF COLMN (ROWIX, ELMCX) NOT = 'N' GO TO DX052.                   DETAB-65
     ADD 1 TO COLIX.                                                  DETAB-65
     GO TO DX051-1.                                                   DETAB-65
 DX051-2.                                                             DETAB-65
     MOVE 'EL001' TO DGOLN.                                           DETAB-65
     MOVE DGOTO TO CDOPR.                                             DETAB-65
     GO TO DX202.                                                     DETAB-65
 DX052.                                                               DETAB-65
     MOVE ROWIX TO ELMRX.                                             DETAB-65
                                                                      DETAB-65
     NOTE  ARE THE REST OF THE ELEMENTS IN THIS COLUMN BLANK.         DETAB-65
                                                                      DETAB-65
 DX052-1.                                                             DETAB-65
     IF ELMRX = NOCON GO TO DX052-2.                                  DETAB-65
     COMPUTE ELMCT = ELMRX + 1.                                       DETAB-65
     IF COLMN (ELMCT, ELMCX) NOT = ' ' GO TO DX201.                   DETAB-65
     ADD 1 TO ELMRX.                                                  DETAB-65
     GO TO DX052-1.                                                   DETAB-65
 DX052-2.                                                             DETAB-65
     IF NCOLS = 1 THEN MOVE ROWIX TO NOCON GO TO DX014.               DETAB-65
     MOVE COLIX TO DUMIX.                                             DETAB-65
     GO TO DX202.                                                     DETAB-65
                                                                      DETAB-65
     NOTE  PUSH LAST-IN-FIRST-OUT LIST.                               DETAB-65
                                                                      DETAB-65
 DX201.                                                               DETAB-65
     MOVE 'DX' TO DGOLB.                                              DETAB-65
     ADD 1 TO LABNO, ADD 1 TO LABIX.                                  DETAB-65
     MOVE LABNO TO DGONO, PDPUL (LABIX).                              DETAB-65
     MOVE DGOTO TO CDOPR.                                             DETAB-65
 DX202.                                                               DETAB-65
     MOVE 1 TO COLIX.                                                 DETAB-65
                                                                      DETAB-65
     NOTE  IS THERE A N OR A BLANK IN THIS ROW.                       DETAB-65
```

```
                                                                       DETAB-65
       DX202-1.                                                        DETAB-65
           IF COLIX GREATER NCOLS GO TO DX202-2.                       DETAB-65
           MOVE ELIMC (COLIX) TO ELMCX.                                DETAB-65
           IF COLMN (ROWIX, ELMCX) NOT = 'Y' GO TO DX053.              DETAB-65
           ADD 1 TO COLIX.                                             DETAB-65
           GO TO DX202-1.                                              DETAB-65
       DX202-2.                                                        DETAB-65
           MOVE 'EL001' TO DGOLN.                                      DETAB-65
           MOVE ' ELSE ' TO DELSE.                                     DETAB-65
           MOVE DGOTO TO ELOPR.                                        DETAB-65
           PERFORM DX204 THRU DX205.                                   DETAB-65
           GO TO DX300.                                                DETAB-65
       DX053.                                                          DETAB-65
           MOVE ROWIX TO ELMRX.                                        DETAB-65
                                                                       DETAB-65
           NOTE  ARE THE REST OF THE ELEMENTS IN THIS COLUMN BLANK.    DETAB-65
                                                                       DETAB-65
       DX053-1.                                                        DETAB-65
           IF ELMRX = NOCON GO TO DX053-2.                             DETAB-65
           COMPUTE ELMCT = 1 + ELMRX.                                  DETAB-65
           IF COLMN (ELMCT, ELMCX) NOT = ' '                           DETAB-65
               MOVE '.' TO PIF, GO TO DX204.                           DETAB-65
           ADD 1 TO ELMRX.                                             DETAB-65
           GO TO DX053-1.                                              DETAB-65
       DX053-2.                                                        DETAB-65
           MOVE ROWIX TO NOCON.                                        DETAB-65
           IF DUMIX NOT = ZERO OR NCOLS = 1 THEN GO TO DX014.          DETAB-65
           MOVE COLIX TO ELMRX.                                        DETAB-65
           MOVE AZ TO DGOLB.                                           DETAB-65
           MOVE ELMCX TO DGONO.                                        DETAB-65
           MOVE ' ELSE ' TO DELSE.                                     DETAB-65
           MOVE DGOTO TO ELOPR.                                        DETAB-65
           PERFORM DX016 THRU DX020.                                   DETAB-65
           PERFORM DX011 THRU DX055.                                   DETAB-65
           MOVE NOCON TO ROWIX.                                        DETAB-65
           MOVE NROWS TO NOCON.                                        DETAB-65
       DX300.                                                          DETAB-65
           MOVE '.' TO PELSE.                                          DETAB-65
           PERFORM L1OUT THRU RITAB.                                   DETAB-65
           PERFORM L2OUT THRU RITAB.                                   DETAB-65
           IF NORLS = ZERO GO TO DX038.                                DETAB-65
           MOVE 'DX' TO LABNM.                                         DETAB-65
           MOVE PDPUL (LABIX) TO LABVL.                                DETAB-65
           SUBTRACT 1 FROM LABIX.                                      DETAB-65
           PERFORM DLOUT THRU RITAB.                                   DETAB-65
           GO TO DX004.                                                DETAB-65
       DX204.                                                          DETAB-65
           IF DUMIX = ZERO GO TO DX205.                                DETAB-65
           MOVE ROWIX TO NOCON.                                        DETAB-65
           MOVE AZ TO DGOLB.                                           DETAB-65
           MOVE ELIMC (DUMIX) TO DGONO.                                DETAB-65
           MOVE DGOTO TO CDOPR.                                        DETAB-65
           MOVE DUMIX TO COLIX.                                        DETAB-65
           PERFORM DX016 THRU DX020.                                   DETAB-65
           MOVE NOCON TO ROWIX.                                        DETAB-65
           MOVE NROWS TO NOCON.                                        DETAB-65
       DX205.                                                          DETAB-65
           EXIT.                                                       DETAB-65
       DX009.                                                          DETAB-65
           PERFORM DX010 THRU DX055 VARYING ELMRX FROM 1 BY 1 UNTIL    DETAB-65
             ELMRX IS GREATER THAN NCOLS.                              DETAB-65
           GO TO DX003.                                                DETAB-65
                                                                       DETAB-65
           NOTE  DELETE FROM PATH INDEX ALL COLUMNS THAT HAVE A Y      DETAB-65
           IN THIS ROW.                                                DETAB-65
```

```
                                                                 DETAB-65
DX010.                                                           DETAB-65
    MOVE ELIMC (ELMRX) TO COLIX.                                 DETAB-65
    IF COLMN (ROWIX, COLIX) NOT = 'Y' GO TO DX055.               DETAB-65
DX011.                                                           DETAB-65
    SUBTRACT 1 FROM NCOLS.                                       DETAB-65
    PERFORM DX012 VARYING ELMCX FROM ELMRX BY 1 UNTIL ELMCX      DETAB-65
      GREATER THAN NCOLS.                                        DETAB-65
    SUBTRACT 1 FROM ELMRX, SUBTRACT 1 FROM COLIX.                DETAB-65
DX012.                                                           DETAB-65
    COMPUTE ELMCT = 1 + ELMCX.                                   DETAB-65
    MOVE ELIMC (ELMCT) TO ELIMC (ELMCX).                         DETAB-65
DX055.                                                           DETAB-65
    EXIT.                                                        DETAB-65
DX014.                                                           DETAB-65
    MOVE ELIMC (1) TO COLIX.                                     DETAB-65
    PERFORM DX015 VARYING ROWIX FROM 1 BY 1 UNTIL ROWIX = NROWS. DETAB-65
DX015.                                                           DETAB-65
    MOVE COLMN (ROWIX, COLIX) TO SAVCL (ROWIX).                  DETAB-65
DX056.                                                           DETAB-65
    MOVE 4 TO DUMIX.                                             DETAB-65
    PERFORM DX022 THRU DX031 VARYING COLIX FROM 1 BY 1 UNTIL     DETAB-65
      COLIX IS GREATER THAN NCOLS.                               DETAB-65
    GO TO DX032.                                                 DETAB-65
                                                                 DETAB-65
    NOTE DETERMINE ACTION LABELS AND CHECK FOR REDUNDENCY.       DETAB-65
                                                                 DETAB-65
DX022.                                                           DETAB-65
    MOVE ELIMC (COLIX) TO ELMCX.                                 DETAB-65
    IF COLMN (NOCON, ELMCX) NOT = 'Y' GO TO DX029.               DETAB-65
    IF DUMIX = 3 OR DUMIX = 1 THEN GO TO DX059.                  DETAB-65
    IF DUMIX = 2 MOVE 3 TO DUMIX ELSE MOVE 1 TO DUMIX.           DETAB-65
    MOVE AZ TO DGOLB.                                            DETAB-65
    MOVE ' ELSE ' TO DELSE.                                      DETAB-65
    MOVE ELMCX TO DGONO.                                         DETAB-65
    MOVE DGOTO TO CDOPR.                                         DETAB-65
    GO TO DX031.                                                 DETAB-65
DX059.                                                           DETAB-65
    MOVE WRNING-2 TO WRNING-IMAGE.                               DETAB-65
    WRITE TAPE-LIST FROM WRNING-PRINT.                           DETAB-65
    PERFORM DX028 VARYING ELMRX FROM 1 BY 1 UNTIL ELMRX = NCOLS. DETAB-65
DX028                                                            DETAB-65
    MOVE           RULE' TO TAPE-LIST.                           DETAB-65
    MOVE ELIMC (ELMRX) TO CARDX.                                 DETAB-65
    WRITE TAPE-LIST.                                             DETAB-65
DX013.                                                           DETAB-65
    EXIT.                                                        DETAB-65
DX029.                                                           DETAB-65
    IF COLMN (NOCON, ELMCX) NOT = 'N' GO TO DX031.               DETAB-65
    IF DUMIX = 3 OR DUMIX = 2 PERFORM DX059 THRU DX013,          DETAB-65
      GO TO DX031.                                               DETAB-65
    IF DUMIX = 1 MOVE 3 TO DUMIX ELSE MOVE 2 TO DUMIX.           DETAB-65
    MOVE AZ TO DGOLB.                                            DETAB-65
    MOVE ' ELSE ' TO DELSE.                                      DETAB-65
    MOVE ELMCX TO DGONO.                                         DETAB-65
    MOVE DGOTO TO ELOPR.                                         DETAB-65
DX031.                                                           DETAB-65
    EXIT.                                                        DETAB-65
DX032.                                                           DETAB-65
    MOVE 'EL001' TO DGOLN.                                       DETAB-65
    MOVE '.' TO PELSE.                                           DETAB-65
    IF DUMIX = 2 MOVE DGOTO TO CDOPR ELSE                        DETAB-65
      IF DUMIX = 1 MOVE DGOTO TO ELOPR.                          DETAB-65
    MOVE CNDI (NOCON) TO COND.                                   DETAB-65
    PERFORM DX016 THRU DX020 VARYING COLIX FROM 1 BY 1 UNTIL     DETAB-65
```

```
         COLIX IS GREATER THAN NCOLS.                                DETAB-65
       GO TO DX301.                                                  DETAB-65
   DX016.                                                            DETAB-65
       MOVE ELIMC (COLIX) TO DUMIX.                                  DETAB-65
       MOVE 1 TO ROWIX.                                              DETAB-65
   DX016-1.                                                          DETAB-65
       IF ROWIX GREATER NOCON GO TO DX016-2.                         DETAB-65
       IF COLMN (ROWIX, DUMIX) = 'B' GO TO DX504.                    DETAB-65
       ADD 1 TO ROWIX.                                               DETAB-65
       GO TO DX016-1.                                                DETAB-65
   DX016-2.                                                          DETAB-65
       MOVE 0 TO ROWIX.                                              DETAB-65
   DX016-3.                                                          DETAB-65
       IF ROWIX = NOCON GO TO DX016-4.                               DETAB-65
       COMPUTE ELMCX = NOCON - ROWIX.                                DETAB-65
       IF COLMN (ELMCX, DUMIX) = ' ' THEN                            DETAB-65
           MOVE 'B' TO COLMN (ELMCX, DUMIX), GO TO DX020.            DETAB-65
       ADD 1 TO ROWIX.                                               DETAB-65
       GO TO DX016-3.                                                DETAB-65
   DX016-4.                                                          DETAB-65
       SUBTRACT 1 FROM NORLS.                                        DETAB-65
       PERFORM DX100 VARYING ELMCX FROM 1 BY 1                       DETAB-65
       UNTIL ELMCX IS GREATER THAN NORLS.                            DETAB-65
       GO TO DX020.                                                  DETAB-65
   DX100.                                                            DETAB-65
       COMPUTE ELMCT = ELMCX + 1                                     DETAB-65
       IF MATIX (ELMCX) IS NOT LESS THAN DUMIX                       DETAB-65
           MOVE MATIX (ELMCT) TO MATIX (ELMCX).                      DETAB-65
   DX504.                                                            DETAB-65
       MOVE 1 TO ELMCT.                                              DETAB-65
   DX504-1.                                                          DETAB-65
       IF ELMCT = ROWIX GO TO DX504-2.                               DETAB-65
       COMPUTE ELMCX = ROWIX - ELMCT.                                DETAB-65
       IF COLMN (ELMCX, DUMIX) = ' ' GO TO DX507.                    DETAB-65
       ADD 1 TO ELMCT.                                               DETAB-65
       GO TO DX504-1.                                                DETAB-65
   DX504-2.                                                          DETAB-65
       MOVE 'Y' TO COLMN (ROWIX, DUMIX).                             DETAB-65
       GO TO DX016.                                                  DETAB-65
   DX507.                                                            DETAB-65
       MOVE 'B' TO COLMN (ELMCX, DUMIX).                             DETAB-65
       PERFORM DX508 VARYING ELMCX FROM ROWIX BY 1                   DETAB-65
          UNTIL ELMCX = NOCON.                                       DETAB-65
       GO TO DX020.                                                  DETAB-65
   DX508.                                                            DETAB-65
       IF COLMN (ELMCX, DUMIX) = 'B'                                 DETAB-65
          MOVE ' ' TO COLMN (ELMCX, DUMIX).                          DETAB-65
   DX020.                                                            DETAB-65
       EXIT.                                                         DETAB-65
                                                                     DETAB-65
       NOTE  POP  LAST-IN-FIRST-OUT LIST.                            DETAB-65
                                                                     DETAB-65
   DX301.                                                            DETAB-65
       PERFORM L1OUT THRU RITAB.                                     DETAB-65
       PERFORM L2OUT THRU RITAB.                                     DETAB-65
       IF NORLS = ZEROES GO TO DX038.                                DETAB-65
       MOVE 'DX' TO LABNM.                                           DETAB-65
       MOVE PDPUL (LABIX) TO LABVL.                                  DETAB-65
       SUBTRACT 1 FROM LABIX.                                        DETAB-65
       PERFORM DLOUT THRU RITAB.                                     DETAB-65
                                                                     DETAB-65
       NOTE  SETUP INDEXES FOR NEXT PASS.                            DETAB-65
                                                                     DETAB-65
   DX302.                                                            DETAB-65
       MOVE NORLS TO NCOLS.                                          DETAB-65
       MOVE NROWS TO NOCON.                                          DETAB-65
```

```
           MOVE HATIT TO ELIMT.                                   DETAB-65
           MOVE 1 TO ROWIX.                                       DETAB-65
       DX302-1.                                                   DETAB-65
           IF ROWIX = NOCON                                       DETAB-65
             MOVE ERR-10 TO ERR-IMAGE                             DETAB-65
             WRITE TAPE-LIST FROM ERR-PRNT                        DETAB-65
             GO TO DT001.                                         DETAB-65
                                                                  DETAB-65
           NOTE  DELETE THAT PATH GENERATED ON THE LAST PASS AND  DETAB-65
           FIND THE NEXT HIGHER NODE ON THE TREE.                 DETAB-65
                                                                  DETAB-65
           MOVE 1 TO COLIX.                                       DETAB-65
       DX034-1.                                                   DETAB-65
           IF SAVCL (ROWIX) = ' ' MOVE 'N' TO SAVCL (ROWIX).      DETAB-65
           IF COLIX GREATER NCOLS GO TO DX004.                    DETAB-65
           MOVE ELIMC (COLIX) TO ELMCX.                           DETAB-65
           IF COLMN (ROWIX, ELMCX) = ' ' OR COLMN (ROWIX, ELMCX)  DETAB-65
               = SAVCL (ROWIX) GO TO DX034-2.                     DETAB-65
           ADD 1 TO COLIX.                                        DETAB-65
           GO TO DX034-1.                                         DETAB-65
       DX034-2.                                                   DETAB-65
           PERFORM DX037 VARYING COLIX FROM 1 BY 1 UNTIL COLIX = NCOLS.  DETAB-65
       DX037.                                                     DETAB-65
           MOVE ELIMC (COLIX) TO ELMCX.                           DETAB-65
           MOVE COLIX TO ELMRX.                                   DETAB-65
           IF COLMN (ROWIX, ELMCX) NOT = ' ' AND COLMN (ROWIX, ELMCX)  DETAB-65
               NOT = SAVCL (ROWIX) PERFORM DX011 THRU DX055.      DETAB-65
       DX061.                                                     DETAB-65
           ADD 1 TO ROWIX.                                        DETAB-65
           GO TO DX302-1.                                         DETAB-65
       DX038.                                                     DETAB-65
           MOVE SPACES TO LINE3.                                  DETAB-65
           COMPUTE KEY-2 = NORULS - 1.                            DETAB-65
           PERFORM DX039 THRU DX0398 VARYING COLIX FROM 1 BY 1    DETAB-65
             UNTIL COLIX = KEY-2.                                 DETAB-65
       DX039.                                                     DETAB-65
           MOVE AZ TO LABNM.                                      DETAB-65
           MOVE COLIX TO LABVL.                                   DETAB-65
           PERFORM DLOUT THRU RITAB.                              DETAB-65
           ADD 1 NCOND GIVING KEY-1.                              DETAB-65
           PERFORM DXA01 THRU DXA04 VARYING EXIX FROM 1 BY 1 UNTIL  DETAB-65
             EXIX IS GREATER THAN NACTS.                          DETAB-65
           MOVE SPACES TO CRD-OUT.                                DETAB-65
           EXAMINE DUM-5 TALLYING UNTIL FIRST 'G'.                DETAB-65
           IF TALLY = 58 GO TO DX039H.                            DETAB-65
           IF TALLY NOT = ZERO, THEN                              DETAB-65
               IF TEMP (TALLY) NOT = SPACE GO TO DX039H.          DETAB-65
           COMPUTE DUMIX = TALLY + 1.                             DETAB-65
           PERFORM DX039F VARYING TALLY FROM 1 BY 1 UNTIL TALLY = 6.  DETAB-65
           GO TO DX039G.                                          DETAB-65
       DX039F.                                                    DETAB-65
           MOVE TEMP (DUMIX) TO EGOTO (TALLY).                    DETAB-65
           ADD 1 TO DUMIX.                                        DETAB-65
       DXA01.                                                     DETAB-65
           IF COLMN (KEY-1, COLIX) = ' ' GO TO DXA04.             DETAB-65
           MOVE ATBL (EXIX) TO DUM-5, BNAME.                      DETAB-65
           PERFORM RSCAN.                                         DETAB-65
           ADD 3 TO DUMIX.                                        DETAB-65
           MOVE '.' TO TCOLS (DUMIX).                             DETAB-65
           PERFORM L2OUT THRU RITAB.                              DETAB-65
       DXA04.                                                     DETAB-65
           ADD 1 TO KEY-1.                                        DETAB-65
       DX039G.                                                    DETAB-65
           IF DUM-4 = 'GO TO' GO TO DX039B.                       DETAB-65
       DX039H.                                                    DETAB-65
           MOVE 'GO TO DEXIT.' TO B-MARG OF CRD-OUT.              DETAB-65
```

```
    MOVE CRD-OUT TO TAPE-LIST.                                        DETAB-65
    PERFORM RITAB.                                                    DETAB-65
DX0398.                                                               DETAB-65
    EXIT.                                                             DETAB-65
DX040.                                                                DETAB-65
    MOVE SPACES TO LINE3.                                             DETAB-65
    COMPUTE KEY-1 = NCOND + 1.                                        DETAB-65
    MOVE NORULS TO COLIX.                                             DETAB-65
    MOVE 1 TO EXIX.                                                   DETAB-65
    MOVE KEY-1 TO TALLY.                                              DETAB-65
    MOVE 0 TO NRLS.                                                   DETAB-65
DX040-2.                                                              DETAB-65
    IF EXIX GREATER NACTS GO TO DX040-3.                              DETAB-65
    IF COLMN (TALLY, COLIX) NOT = ' ' ADD 1 TO NRLS.                  DETAB-65
    ADD 1 TO TALLY, ADD 1 TO EXIX.                                    DETAB-65
    GO TO DX040-2.                                                    DETAB-65
DX040-3.                                                              DETAB-65
    IF NRLS = ZEROES GO TO DX040-1.                                   DETAB-65
    MOVE 'EL001' TO DUM-6.                                            DETAB-65
    PERFORM DLOUT THRU RITAB.                                         DETAB-65
DX040-1.                                                              DETAB-65
    PERFORM DXA01 THRU DXA04 VARYING EXIX FROM 1 BY 1 UNTIL           DETAB-65
      EXIX IS GREATER THAN NACTS.                                     DETAB-65
    MOVE SPACES TO CRD-OUT.                                           DETAB-65
    MOVE 'DEXIT. EXIT.' TO BODY OF CRD-OUT.                           DETAB-65
    MOVE CRD-OUT TO TAPE-LIST.                                        DETAB-65
    PERFORM RITAB.                                                    DETAB-65
    GO TO DT001.                                                      DETAB-65
L1OUT.                                                                DETAB-65
    MOVE LINE1  TO CRD-OUT, TAPE-LIST.  GO TO RITAB.                  DETAB-65
L2OUT.                                                                DETAB-65
    MOVE LINE2  TO CRD-OUT, TAPE-LIST.  GO TO RITAB.                  DETAB-65
DLOUT.                                                                DETAB-65
    MOVE DLABEL TO CRD-OUT, TAPE-LIST.                                DETAB-65
RITAB.                                                                DETAB-65
    WRITE TAPE-LIST.                                                  DETAB-65
    WRITE CRD-OUT.                                                    DETAB-65
RSCAN.                                                                DETAB-65
    MOVE 58 TO DUMIX.                                                 DETAB-65
    PERFORM RS001 THRU RS003.                                         DETAB-65
RS001.                                                                DETAB-65
    IF TEMP (DUMIX) = SPACE GO TO RS002.                              DETAB-65
    ADD 2 TO DUMIX.                                                   DETAB-65
    GO TO RS003.                                                      DETAB-65
RS002.                                                                DETAB-65
    IF DUMIX = 1 GO TO RS003.                                         DETAB-65
    SUBTRACT 1 FROM DUMIX.                                            DETAB-65
    GO TO RS001.                                                      DETAB-65
RS003.                                                                DETAB-65
    EXIT.                                                             DETAB-65
                                                                      DETAB-65
    NOTE DIAGNOSTIC SECTION.                                          DETAB-65
                                                                      DETAB-65
EM01.                                                                 DETAB-65
    MOVE ERR-1 TO ERR-IMAGE.                                          DETAB-65
    GO TO EM99.                                                       DETAB-65
EM02.                                                                 DETAB-65
    MOVE ERR-2 TO ERR-IMAGE.                                          DETAB-65
    GO TO EM99.                                                       DETAB-65
EM03.                                                                 DETAB-65
    MOVE ERR-3 TO ERR-IMAGE.                                          DETAB-65
    GO TO EM99.                                                       DETAB-65
EM04.                                                                 DETAB-65
    MOVE ERR-4 TO ERR-IMAGE.                                          DETAB-65
    GO TO EM99.                                                       DETAB-65
EM05.                                                                 DETAB-65
```

```
       MOVE ERR-5 TO ERR-IMAGE.                                    DETAB-65
       GO TO EM99.                                                 DETAB-65
   EM06.                                                           DETAB-65
       MOVE ERR-6 TO ERR-IMAGE.                                    DETAB-65
       GO TO EM99.                                                 DETAB-65
   EM07.                                                           DETAB-65
       MOVE ERR-7 TO ERR-IMAGE.                                    DETAB-65
       GO TO EM99.                                                 DETAB-65
   EM08.                                                           DETAB-65
       MOVE ERR-8 TO ERR-IMAGE.                                    DETAB-65
       GO TO EM99.                                                 DETAB-65
   EM09.                                                           DETAB-65
       MOVE ERR-9 TO ERR-IMAGE.                                    DETAB-65
   EM99.                                                           DETAB-65
       WRITE TAPE-LIST FROM ERR-PRNT.                              DETAB-65
   READ-1.                                                         DETAB-65
       READ CARD-INPUT INTO DETAB-CRD, AT END GO TO EOF.           DETAB-65
       MOVE SPACES TO IDFLD.                                       DETAB-65
       IF IDENT OF DETAB-CRD = '0000'                              DETAB-65
         MOVE '0' TO TAPE-LIST,                                    DETAB-65
         WRITE TAPE-LIST.                                          DETAB-65
       WRITE TAPE-LIST FROM DETAB-CRD.                             DETAB-65
       IF IDENT OF DETAB-CRD = '999X' GO TO EOF.                   DETAB-65
   SKIP01.                                                         DETAB-65
       IF LINE-ID OF DETAB-CRD NOT = 'S' GO TO READ-1.             DETAB-65
       GO TO DT001.                                                DETAB-65
   EOF.                                                            DETAB-65
       MOVE '0END DETAB/65 PREPROCESSOR RUN.' TO TAPE-LIST.        DETAB-65
       WRITE TAPE-LIST.                                            DETAB-65
       CLOSE CARD-INPUT WITH LOCK.                                 DETAB-65
       CLOSE CARD-OUTPUT WITH LOCK, LIST-OUTPUT WITH LOCK.         DETAB-65
       STOP RUN.                                                   DETAB-65
$STOP
```

Detab/65 User's Manual

1. INTRODUCTION

Decision Tables endow a user with the ability to provide a graphical representation of a complex procedure in such a way that one individual is able to readily understand a program written by another.

DETAB/65 is the decision table language which the preprocessor converts to COBOL statements for subsequent processing by an appropriate COBOL compiler. This manual's purpose is to describe to a user how a DETAB/65 decision table should be written for inclusion within a COBOL program. It also describes the necessary linkages, formats, and restrictions used in construction of the decision table.

Since not all terms will be defined in this manual, it is recommended that the user first study the DETAB/65 documents accompanying this manual.

2. STRUCTURE OF A DECISION TABLE

A decision table can be logically divided into four sections (See Fig. 18-3). The upper two sections (Condition Stub and Condition Entry) describe the set or string of conditions that is to be tested. The lower two sections (Action Stub and Action Entry) describe the set or string of actions that is to be taken upon satisfaction of a set of conditions. A *rule* consists of a set of conditions plus a set of actions, and a *decision table* typically consists of several rules.

Condition Stub	Condition Entry
Action Stub	Action Entry

FIG. 18-3

The three types of decision tables in current use today are the limited-, extended-, and mixed-entry types (see Fig. 18-4). Eventually it will be possible to convert all three types of tables via the preprocessor; however, at the present time the preprocessor is *restricted* to *limited-entry tables.*

	R_1	R_2		R_1	R_2		R_1	R_2
C_1	N	Y	C_1	$= 58$	$=25$	C_1	$= 58$	Y
C_2	Y	N	C_2	$\neq J$	$= K$	C_2	$\neq J$	N
A_1	X	—	A_1	X	—	A_1	X	—
A_2	—	X	A_2	—	X	A_2	—	X
LIMITED-ENTRY			EXTENDED-ENTRY			MIXED-ENTRY		

FIG. 18-4

3. PROGRAM FORMAT

The format for a COBOL program containing DETAB/65 decision tables must conform to the requirements for any COBOL program, except that a decision table is inserted in the COBOL PROCEDURE DIVISION as a SECTION, and is referred to by an appropriate COBOL statement (see Table Linkage).

The DATA DIVISION of a COBOL program incorporating a DETAB/65 decision table is treated as in any other COBOL DATA DIVISION. Any symbolic data reference, data structure, constant, or working storage used in a decision table must be declared in the DATA DIVISION.

The decision table(s) is placed at the end of the PROCEDURE DIVISION since it is to be treated as a COBOL subroutine. This is the only difference between a COBOL program and a COBOL with a decision-table(s) program.

4. TABLE LINKAGE

In compilation a decision table will be treated as a closed COBOL subroutine. Thus, a decision table should not be entered via the normal operating sequence, but only by using COBOL GO TO or PERFORM verbs.*

Since a GO TO results in an uncondition transfer, a return or transfer point must be specified by the user in the decision tables action stub or the processing sequence will be lost. It is recommended that the GO TO verb not be used when referring to a table from the main sequence of the program. When transferring control to a decision table by the use of a COBOL PERFORM verb, a normal return to the processing sequence will be made by the compiler unless the user specifies otherwise in his actions. Specifically, the preprocessor will generate a "GO TO DEXIT" for every rule whose last action does not end in a "GO TO," no matter how the table was entered.

Tables may be chained together by placing GO TO's and PERFORM's in the Action Stub of one or more tables. However, it is advisable to keep very, very close track of this as it is possible to generate errors due to the way various COBOL compilers set up their procedure sections.

5. DEFINITIONS

The following is a series of decision table definitions to be used in describing a DETAB/65 decision table:

* It is possible to enter a table from the main sequence, but the trouble this can cause does not make it worthwhile.

5.1 TABLE-ID

Notifies the preprocessor that a DETAB/65 decision table has been encountered. The ID is always 4 numeric characters consisting of 4 zeros (0000).

5.2. RULE-ID

Notifies the preprocessor that the rule-card is present (for error checking purposes). The ID is always 4 numeric characters in length and consists of (0001).

5.3 TABLE-NAME

This is a 30 alphanumeric character or less name which is then used to identify the COBOL section generated by the preprocessor.

5.4 FORM

Designates the kind of table or present (i.e., limited-, extended-, mixed-entry) and is always an alphabetic character (L, E, M, left-justified).

5.5. COND ROWS

Designates the number of conditions in the Condition Stub of the table. This is 3 numeric characters (right-justified).

5.6 ACTION ROWS

Designates the number of actions in the Action Stub of the table. This is 3 numeric characters (right-justified).

5.7 RULES

Designates the number of rules in a table. This includes the ELSE-RULE and is 3 numeric characters (right-justified).

5.8 RULE NUMBERS

These are 3 numeric characters each used to identify each rule. The ELSE-RULE is the only exception and is designated by the 3 alphabetic characters ELS.

5.9 CONDITION STUB

Contains logical, arithmetic, or relational conditions answerable by a yes (Y) or a no (N).

5.10 CONDITION ENTRIES

These indicate which condition must be met to satisfy a rule. This can be a Y, N, BLANK ( ), or DASH (-). A blank or dash means that the user does not care if the rule is Y or N as it makes no difference. Also known as the elements of a task.

5.11 ACTION STUB

Contains imperative statements to be performed as indicated by the ACTION ENTRIES of a rule when the rule is satisfied.

5.12 ACTION ENTRIES

These indicate which actions must be performed if a rule is satisfied. The character used to signify this is an X.

6. CONVENTIONS AND RESTRICTIONS

When writing DETAB/65 decision tables the following words (6.1–6.6) cannot be used (this in addition to COBOL restricted words).

6.1 DXN (where N is a 3 digit number)
6.2 AZM (where M is a 3 digit number)
6.3 AZP (where P is a 3 digit number)
6.4 EL001
6.5 DEXIT
6.6 ELS
6.7 Maximum of 50 entries in Condition Stub.
6.8 Maximum of 50 entries in Action Stub.
6.9 Maximum of 50 rules (includes ELSE-RULE).
6.10 Maximum of 12 and minimum of 3 columns in a rule.
6.11 Maximum of 58 and minimum of 12 columns in the Condition & Action Stubs.
6.12 ELSE-RULE (ELS) must always be present.
6.13 TABLE END card ($ in Column 7) must always be present.
6.14 EOF card (999X) in Columns 4-7 must always be present.

7. DETAB/65 DECISION TABLE

The table itself is written in a mixture of fixed and free formats. The following lists the various sections showing the proper way to set up a DETAB/65 decision table.

7.1 HEADER

The header contains information used by the preprocessor to initiate the conversion of the table and allow it to check for errors.

COLUMNS(S)	DEFINITION
4-7	TABLE-ID
9-38	TABLE NAME
39-40	FORM
41-43	COND-ROWS
44-46	ACTION-ROWS
47-49	RULES

All other columns in the header are blank.

7.2 RULE

The rule part of the DETAB/65 table contains information which allows the preprocessor to determine the size of the stub area. The size of the largest entry in either the condition or action stub determines the size of the stub area.

COLUMN(S)	DEFINITION
4-7	RULE-ID
9-XX	Blank. This is the size of the stub area.

The rules (RULE-NUMBERS) start immediately after the last column in the stub area and are terminated by a dollar sign ($). The only exception is if the last rule occupies column 80, thus putting the dollar sign in column 81.

7.3 BODY

The body part of the table contains the Condition and Action Stubs and Entries. All actions are performed in the order of their occurrence; if it is desired to do series of actions in a different order, it is necessary to repeat them in that order in the action stub.

7.4 END

The end part notifies the preprocessor when the input is completed and conversion is to begin. It consists of a dollar sign ($) in column 7.

Preprocessor Output. The preprocessor will convert a DETAB/65 decision table into COBOL statements of the following format:

$$\text{IF } C_n \text{ GO TO } \begin{bmatrix} \text{DXN} \\ \text{AXM} \\ \text{EL001} \end{bmatrix} \text{ [.] ELSE GO TO } \left[\begin{bmatrix} \text{AZP} \\ \text{EL001} \end{bmatrix} \right]$$

Where C_n is the condition to be tested, and the brackets are output formats.

Errors. There are two major types of errors encountered with use of a DETAB/65 decision table. These have to do with redundant and contradictory rules and proper use of "OR" relationships. The following is a more detailed description.

Any two or more rules will be either redundant or contradictory if they do not have at least one (Y) (N) combination in at least one condition row. For example, in Fig. 18-5 all 10 of the rules are either redundant or contradictory to each other.

Rules are redundant if the actions are the same for the rules involved, otherwise they are contradictory.

FIG. 18-5

1	2	3	4	5	6	7	8	9	10
Y	Y	—	—	—	Y	Y	Y	Y	Y
Y	Y	Y	—	—	Y	—	—	Y	Y
—	N	N	—	N	—	—	—	N	N
N	N	—	N	N	—	—	N	—	—

An OR relationship is one in which the following occurs: 'If C1 = 1 or C1 = 2 *AND* C2 = 4 *AND* C2 = 4 *AND* C3 = 5 *THEN* . . . .' For the OR's to be properly handled by the preprocessor the decision table must be set up as:

C1 = 1	Y	N
C2 = 2	—	Y
C3 = 4	Y	Y
C4 = 5	Y	Y

The reason for 2 rules is that the above statement is really 2 statements, 'IF C1 = 1 *AND* C2 = 4 *AND* C3 = 5 *THEN* . . .', and IF C1 = 2 *AND* C2 = 4 *AND* C3 = 5 *THEN* . . .'. Since the conditions are handled sequentially from top to bottom, it is not necessary to check for C1 = 2 in rule 1 (if C1 = 1 it cannot be equal to 2): however, in rule 2 C1 = 1 must be checked before C1 = 2 is and thus C1 = 1 must have an 'N' in rule 2. The principle can be extended to any number of "OR" relationships.

8. EOF CARD

To tell the preprocessor when to terminate operations a special card is used called the EOF card. This consists of a 999X in columns 4-7, and is the last card processed. This means that if more than one program is to be processed the EOF card is placed after the last program.

Deck Outline. The following shows the deck structure of a COBOL program containing a DETAB/65 decision table.

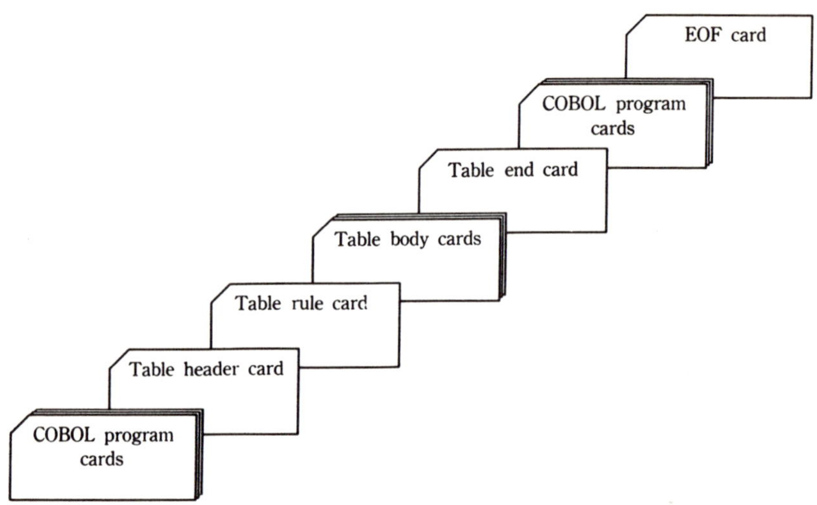

FIG. 18-6. SAMPLE COBOL, DETAB/65 PROGRAM DECK

Appendix A

1. PRESENTLY TABLE RESTRICTED TO LIMITED ENTRIES
Form does not contain an 'L'.
2. TABLE-NAME MISSING FROM HEADER CARD
Processing will be halted and table skipped.
3. RULES CARD MISSING
Processing will be halted and table skipped.
4. LESS THAN 3 RULE COLUMNS SPECIFIED
A rule contains 2 or less columns for processing, conversion will be halted and table skipped.
5. PRESENTLY, CONTINUED RULES NOT IMPLEMENTED
Column 8 has an entry.
6. CONDITION STUB EXCEEDS 58 COLUMNS
Condition stub exceeds limit processing halted and table skipped.
7. NUMBER OF RULES ENCOUNTERED DISAGREES WITH RULE CARD
Number of rules entered in header card (RULES) differs from that specified by the user.
8. MORE THAN 50 ACTION OR CONDITION ENTRIES
Action or Condition stub contains more than 50 entries, processing halted and table skipped.
9. DECISION TABLE LOGIC ERROR. PROCESSING HALTED.
Check over rules for either redundancy, inconsistency or both.

```
+BEGIN JOB 296  08/10/65
+COOP,90101,DETAB65,8/1S,56,55555,4.
+COBOL,X.
       IDENTIFICATION DIVISION.
       PROGRAM-ID,  PLAYBOY.
       AUTHOR,  CHARLES CREE.
       DATE-COMPLIED.  TODAY.
       REMARKS,  THIS IS A SIMPLE DATE RETRIEVAL PROGRAM TO
           ILLUSTRATE THE USE OF DETAB/65.
       ENVIRONMENT DIVISION.
       CONFIGURATION SECTION.
       SOURCE-COMPUTER.  CONTROL DATA 1604-A.
       OBJECT-COMPUTER.  CONTROL DATA 1604-A.
       INPUT-OUTPUT SECTION.
       FILE-CONTROL.
           SELECT CANDIDATES ASSIGN TO SYSTEM-INPUT-TAPE.
           SELECT RATED-FILE ASSIGN TO SYSTEM-OUTPUT-TAPE.
       DATA DIVISION.
       FILE SECTION.
       FD CANDIDATES
           LABEL RECORDS ARE OMITTED
           DATA RECORD IS INREC.
       01  INREC.
           02 FILLER  PICTURE X(80).
       FD  RATED-FILE
           LABEL RECORDS ARE OMITTED
           DATA RECORD IS EVAL.
       01  EVAL.
           02 FILLER  PICTURE X(48).
       WORKING-STORAGE SECTION.
       77  TOP-CTR            PICTURE 9(7)  COMPUTATIONAL,
             SYNCHRONIZED RIGHT,  VALUE ZERO,
       77  NEXT-BEST-CTR      PICTURE 9(7)  COMPUTATIONAL,
             SYNCHRONIZED RIGHT,  VALUE ZERO.
       77  LAST-RESORT-CTR  PICTURE 9(7)  COMPUTATIONAL,
             SYNCHRONIZED RIGHT,  VALUE ZERO.
       77  TOTAL-NO PICTURE 9(5) COMPUTATIONAL,
             SYNCHRONIZED RIGHT, VALUE ZERO.
       01  WSR.
        02  SKIP-CTL PICTURE X,
        02  RATING   PICTURE X(12),
        02  WSR1.
        03  IDNO     PICTURE 9(7).
        03  SEX      PICTURE X.
          88 FEMALE   VALUE ≠F≠.
        03  AGE      PICTURE 999.
        03  HEIGHT   PICTURE 999.
        03  WEIGHT   PICTURE 999.
        03  HAIR     PICTURE X(9).
          88 BLOND    VALUE ≠BLOND≠.
        03  EYES     PICTURE X(5),
          88 BLUE-EYES VALUE ≠BLUE≠.
        03  I-Q      PICTURE 999
        03  FILLER   PICTURE X.
        PROCEDURE DIVISION.
        PNO1.
            OPEN INPUT CANDIDATES.
            OPEN OUTPUT RATED-FILE.
        PNO2.
            READ CANDIDATES INTO WSR1,  AT END GO TO PNEOJ.
            PERFORM CHOICE-PICK.
            GO TO PNO2.
```

```
PNEOJ.
    CLOSE CANDIDATES WITH LOCK, RATED-FILE WITH LOCK.
    DISPLAY ≠TOPS COUNT = ≠ TOP-CTR.
    DISPLAY ≠NEXT COUNT = ≠ NEXT-BEST-CTR.
    DISPLAY ≠LAST COUNT = ≠ LAST-RESORT-CTR.
    DISPLAY ≠TOTAL = ≠ TOTAL-NO.
    STOP RUN.
0010000 CHOICE-PICK                         L 010008008
 010001                                     00100200300400500600 7ELSs
 010101 FEMALE                               Y   Y   Y   Y   Y   Y   Y
 010102 AGE GREATER THAN 18                  Y   Y   Y   Y   Y   Y   Y
 010103 AGE LESS THAN 38                     Y   N   Y   Y   Y  .N   Y
 010104 BLOND                                Y       N           N
 010105 BLUE-EYES                            Y       N       N   N
 010106 WEIGHT GREATER THAN 89               Y   Y   Y   Y   Y   Y   Y
 010107 WEIGHT LESS THAN 133                 Y   Y   Y   N   Y   N   Y
 010108 HEIGHT GREATER THAN 59               Y   Y   Y   N   Y   N   Y
 010109 HEIGHT LESS THAN 68                  Y   Y   Y       Y       N
 010110 I-Q GREATER THAN 99                      Y   N   Y   Y   N   N

 010112 MOVE ≠TOPS≠ TO RATING                X               X   X
 010202 ADD 1 TO TOP-CTR                     X               X   X
 010203 MOVE ≠NEXT BEST≠ TO RATING                   X               X
 010204 ADD 1 TO NEXT-BEST-CTR                       X               X
 010204 MOVE ≠LAST RESORT≠ TO RATING             X       X
 010206 ADD 1 TO LAST-RESORT-CTR                 X       X
 010201 WRITE EVAL FROM WSR                  X   X   X   X   X   X   X
 010207 ADD 1 TO TOTAL-NO                    X   X   X   X   X   X   X   X
     $

      CHOICE-PICK SECTION.
      DX000.
          IF FEMALE
           GO TO DX001 ELSE GO TO EL001.
      DX001.
          IF AGE GREATER THAN 18
           GO TO DX002 ELSE GO TO EL001.
      DX002.
          IF AGE LESS THAN 38
           GO TO DX003.
          IF BLOND
           GO TO DX004.
          IF BLUE-EYES
           GO TO DX005.
          IF WEIGHT GREATER THAN 89
           GO TO DX006 ELSE GO TO EL001.
      DX006.
          IF WEIGHT LESS THAN 133
  GO TO DX007.
  IF HEIGHT GREATER THAN 59
   GO TO EL001.
  IF I-Q GREATER THAN 99
   GO TO EL001 ELSE GO TO AZ006.
DX007.
    IF HEIGHT GREATER THAN 59
     GO TO DX008 ELSE GO TO EL001.
DX008.
    IF HEIGHT LESS THAN 68
     GO TO DX009 ELSE GO TO EL001.
DX009.
    IF I-Q GREATER THAN 99
     GO TO AZ002 ELSE GO TO EL001.
DX005.
    IF WEIGHT GREATER THAN 89
     GO TO DX010 ELSE GO TO EL001.
```

```
DX010.
    IF WEIGHT LESS THAN 133
     GO TO DX011 ELSE GO TO EL001.
DX011.
    IF HEIGHT GREATER THAN59
     GO TO DX012 ELSE GO TO EL001.
DX012.
    IF HEIGHT LESS THAN 68
     GO TO DX013 ELSE GO TO EL001.
DX013.
    IF I-Q GREATER THAN 99
     GO TO AZ002 ELSE GO TO EL001.
DX004.
    IF WEIGHT GREATER THAN 89
     GO TO DX014 ELSE GO TO EL001.
DX014.
    IF WEIGHT LESS THAN 133
     GO TO DX015 ELSE GO TO EL001.
DX015.
    IF HEIGHT GREATER THAN 59
     GO TO DX016 ELSE GO TO EL001.
DX016.
    IF HEIGHT LESS THAN 68
     GO TO DX017 ELSE GO TO EL001.
DX017.
    IF I-Q GREATER THAN 99
     GO TO AZ002 ELSE GO TO EL001.
DX003.
    IF BLOND
     GO TO DX018.
    IF BLUE-EYES.
     GO TO DX019.
    IF WEIGHT GREATER THAN 89
     GO TO DX020 ELSE GO TO EL001.
DX020.
    IF WEIGHT LESS THAN 133
     GO TO DX021.
    IF HEIGHT GREATER THAN 59
     GO TO EL001.
    IF I-Q GREATER THAN 99
     GO TO AZ004 ELSE GO TO EL001.
DX021.
    IF HEIGHT GREATER THAN 59
     GO TO DX022 ELSE GO TO EL001.
DX022.
    IF HEIGHT LESS THAN 68
     GO TO DX023.
    IF I-Q GREATER THAN 99
     GO TO EL001 ELSE GO TO AZ007.
DX023.
    IF I-Q GREATER THAN 99
     GO TO AZ005 ELSE GO TO AZ003.
DX019.
    IF WEIGHT GREATER THAN 89
     GO TO DX024 ELSE GO TO EL001.
DX024.
    IF WEIGHT LESS THAN 133
     GO TO DX025.
    IF HEIGHT GREATER THAN 59
     GO TO EL001.
    IF I-Q GREATER THAN 99
     GO TO AZ004 ELSE GO TO EL001.
DX025.
    IF HEIGHT GREATER THAN 59
     GO TO DX026 ELSE GO TO EL001.
DX026.
     GO TO EL001.
    IF I-Q GREATER THAN 99
     GO TO EL001 ELSE GO TO AZ007.
DX018.
    IF BLUE-EYES
     GO TO DX027.
    IF WEIGHT GREATER THAN 89
     GO TO DX028 ELSE GO TO EL001.
DX028.
    IF WEIGHT LESS THAN 133
     GO TO DX029.
    IF HEIGHT GREATER THAN 59
     GO TO EL001.
    IF I-Q GREATER THAN 99
     GO TO AZ004 ELSE GO TO EL001.
DX029.
    IF HEIGHT GREATER THAN 59
     GO TO DX030 ELSE GO TO EL001.
DX030.
    IF HEIGHT LESS THAN 68
     GO TO DX031.
    IF I-Q GREATER THAN 99
     GO TO EL001 ELSE GO TO AZ007.
DX031.
    IF I-Q GREATER THAN 99
     GO TO AZ005 ELSE GO TO EL001.
DX027.
    IF BLUE-EYES
     GO TO DX032.
    IF WEIGHT GREATER THAN 89
     GO TO DX033 ELSE GO TO EL001.
DX033.
    IF WEIGHT LESS THAN 133
     GO TO DX034.
    IF HEIGHT GREATER THAN 59
     GO TO EL001.
    IF I-Q GREATER THAN 99
     GO TO AZ004 ELSE GO TO EL001.
DX034.
    IF HEIGHT GREATER THAN 59
     GO TO DX035 ELSE GO TO EL001.
DX035.
    IF HEIGHT LESS THAN 68
     GO TO EL001.
    IF I-Q GREATER THAN 99
     GO TO EL001 ELSE GO TO AZ007.
DX032.
    IF WEIGHT GREATER THAN 89
     GO TO DX036 ELSE GO TO EL001.
DX036.
    IF WEIGHT LESS THAN 133
     GO TO DX037 ELSE GO TO EL001.
DX037.
    IF HEIGHT GREATER THAN 59
     GO TO DX038 ELSE GO TO EL001.
DX038.
    IF HEIGHT LESS THAN 68
     GO TO AZ001 ELSE GO TO EL001.
AZ001.
    MOVE ≠TOPS≠ TO RATING.
    ADD 1 TO TOP-CTR.
    WRITE EVAL FROM WSR.
    ADD 1 TO TOTAL-NO.
    GO TO DEXIT.
AZ002.
    MOVE ≠LAST RESORT≠ TO RATING.
```

```
    ADD 1TO LAST-RESORT-CTR.
    WRITE EVAL FROM WSR.
    ADD 1 TO TOTAL-NO.
    GO TO DEXIT.
AZ003.
    MOVE ≠NEXT BEST≠ TO RATING.
    ADD 1 TO NEXT-BEST-CTR.
    WRITE EVAL FROM WSR.
    ADD 1 TO TOTAL-NO.
    GO TO DEXIT.
AZ004.
    MOVE ≠LAST RESORT≠ TO RATING.
    ADD 1 TO LAST-RESORT-CTR.
    WRITE EVAL FROM WSR.
    ADD 1 TO TOTAL-NO.
    GO TO DEXIT.
AZ005.
    MOVE ≠TOPS≠ TO RATING.
    ADD 1 TO TOP-CTR.
    WRITE EVAL FROM WSR.
    ADD 1 TO TOTAL-NO.
    GO TO DEXIT.
AZ006.
    MOVE ≠TOPS≠ TO RATING.
    ADD 1 TO TOP-CTR.
    WRITE EVAL FROM WSR.
    ADD 1 TO TOTAL-NO.
    GO TO DEXIT.
AZ007.
    MOVE ≠NEXT BEST≠ TO RATING.
    ADD 1 TO NEXT-BEST-CTR.
    WRITE EVAL FROM WSR.
    ADD 1 TO TOTAL-NO.
    GO TO DEXIT.
EL001.
    ADD 1 TO TOTAL-NO.
DEXIT. EXIT.
       END PROGRAM.
   999X
END DETAB/65 PREPROCESSOR RUN.
```

```
         3278215F031067115BLACK     GREEN106
         0567982F025065119BLOND     BLUE 103
         1425893F021062110RED       GREY 101
         0053247050055145BALD       BROWN 98

TOPS     3278215F031067115BLACK     GREEN106
TOPS     1425893F021062110RED       GREY 101
TOPS     0567982F025065119BLOND     BLUE 103

TOPS COUNT = 0000003
NEXT COUNT = 0000000
LAST COUNT = 0000000
TOTAL = 00004
```

```
          0010000 CHOICE-PICK                      L 010008008
           010001                                  001002003004005006007ELSs
           010101 FEMALE                            Y  Y  Y  Y  Y  Y  Y
           010102 AGE GREATER THAN 18               Y  Y  Y  Y  Y  Y  Y
           010103 AGE LESS THAN 38                  Y  N  Y  Y  Y  N  Y
           010104 BLOND                             Y     N        N
           010105 BLUE-EYES                         Y     N     N  N
           010106 WEIGHT GREATER THAN 89            Y  Y  Y  Y  Y  Y  Y
           010107 WEIGHT LESS THAN 133              Y  Y  Y  N  Y  N  Y
           010108 HEIGHT GREATER THAN 59            Y  Y  Y  N  Y  N  Y
           010109 HEIGHT LESS THAN 68               Y  Y  Y     Y     N
           010110 I-Q GREATER THAN 99                  Y  N  Y  Y  N  N

           010112 MOVE ≠TOPS≠ TO RATING             X           X  X
           010202 ADD 1 TO TOP-CTR                  X           X  X
           010203 MOVE ≠NEXT BEST≠ TO RATING              X           X
           010204 ADD 1 TO NEXT-BEST-CTR                  X           X
           010204 MOVE ≠LAST RESORT≠ TO RATING         X     X
           010206 ADD 1 TO LAST-RESORT-CTR             X     X
           010201 WRITE EVAL FROM WSR               X  X  X  X  X  X  X
           010207 ADD 1 TO TOTAL-NO                 X  X  X  X  X  X  X  X
                $
                 END PROGRAM.
             999X
```

This article, by SOLOMON L. POLLACK, is from the Systems and Procedures Association's publication, *Ideas for Management—1966*. It is used here by permission.

19

Specific Applications of Decision Tables

Following is a partial list of specific applications of decision tables. This list is reprinted through the courtesy of the Systems & Procedures Association, Cleveland. Ohio.

Accounting

	Analysis	Procedure	Program Documentation
Accounts Payable	x	x	
Accounts Receivable	x	x	x
Accounts Receivable Statements		x	
Analysis of Loan Status Cards	x	x	
Analysis of Vendors	x		
Annuity Classification	x		
Applying Construction Charges		x	
Appropriation Procedures	x	x	
Assignment of Commission Credit	x		

	Analysis	Procedure	Documentation Program
Automatic Premium Loan Repayment	x		
Billing and Billing Adjustments	x	x	x
Billing Brackets			x
Brokerage Calculations	x		
Budget Program			x
Building Standard Code	x		x
Capital Equipment Projects	x		
Capital Stock		x	
Cash Application	x	x	
Cash Disbursements	x	x	

	Analysis	Procedure	Program Documentation
Chart of Accounts Validity Check		x	
Check Writing	x	x	
Collection Follow-Up	x		
Compute Service Charges	x		
Commercial Loan Accounting	x		
Commission Accounting	x		
Computer Audit of Invoices (Frt Application)		x	
Cost Accounting		x	x
Cost Analysis	x	x	
Cost Estimating		x	
Cost File Maintenance	x		
Credit Analysis & Rating	x		
Credit Letter Writing		x	
Dealer Financing	x		
Demand Deposits		x	x
Depreciation—Book and Tax		x	
Determination of Freight Allowances		x	
Discount Determination	x		
Effective Rate Table Maintenance		x	
Escrow Account Analysis	x		
Expense Allocations	x	x	x
Expense Distribution Edit		x	

	Analysis	Procedure	Program Documentation
Expense Ledger	x		
Financial Analysis	x		
Financial Operating Plan	x		x
Freight Demurrage Calculations	x	x	
Freight Rates Calculations		x	
General Accounting			x
General Ledger	x		
Government Bond Purchase		x	
Handling Fee Reports			x
Hospital Accounting	x		x
Interest on Loans		x	
Installment Loans	x	x	x
Insurance Claims (input)	x	x	
Insurance Rates	x	x	
Insurance Rate Formula		x	
Investments	x	x	
Job Cost	x		
Job Cost Distribution	x	x	
Journal Converted to General Accounts	x	x	
Labor Reporting	x		
Labor Standards	x		
Liabilities at Term		x	
Master Vendor No. Assignment	x		
Masterial Accounting	x	x	
Merged Accountability & Fund Reporting	x	x	

	Analysis	Procedure	Program Documentation
MICR Check Reconciliation		x	
Natural Gas Accounting	x	x	
Overhead Analysis	x		
Pension Fund			x
Preparation of Quarterly Tape for Social Security	x		
Pre-Underwriting			x
Product Costs		x	
Production Accounting	x		
Product Tax Rebating		x	
Property Cost Control	x		
Real Estate Billing		x	
Retired Pay System		x	
Revenue Accounting	x	x	
Revenue Edit—Price Checking	x	x	
Route Accounting			x
Royalty Payments		x	
Sales Accounting	x		
Salesmen's Commission	x	x	
Savings	x		

	Analysis	Procedure	Program Documentation
Servicemen's Savings Deposit System		x	
Setting Up Customer Accounts		x	
Stock Dividend Calculation	x		
Stock Purchase Plan	x		
Stores Accounting	x		
Tenant Billing	x		
Territorial Distribution of Monies			x
Test Level of Authorization for Credit Memos		x	
Timekeeping/Payroll/Taxes	x	x	x
Trust Accounting	x	x	
Use of Transaction Codes in a Billing Operation		x	
W-2 Requirements	x		
Work-in-Process Accounting	x		
Work Standards		x	

Shipping

Decision Table applications in this area of activity include:

	Analysis	Procedure	Program Documentation
Freight Decisions		x	
Freight Rates		x	
Shipping Dept. Instructions for Filling Out Shipment Cards		x	
Shipping of Goods		x	
Shipment Matrix	x	x	x
Tonnage Report		x	
Truck Routings		x	

Material Control

Decision Table applications in this area of activity include:

	Analysis	Procedure	Program Documentation
Classification of Inventory	x		
Critical Stock Ratios		x	
Ingot Selection	x		
Inventory Allocation	x		x
Inventory Analysis			x
Inventory Control		x	x
Inventory Control—Explanation of Printed Reports	x		x
Inventory Control Ordering		x	
Inventory Forecasting	x		
Inventory Levels of Material & Supplies	x	x	x
Inventory Movement	x	x	
Inventory Masterfile Maintenance			x
Inventory Order vs. Record Update	x	x	
Inventory Simulator		x	x
Inventory Transaction Codes	x		x
Make or Buy	x		
Order Quantity Calculation			x

	Analysis	Procedure	Program Documentation
Parts List Format Selection		x	
Parts Processing	x		
Processing Inventory Transactions			x
Production Inventory Planning	x		
Retail Stores Validation	x		
Sales Classification of Inventory	x		

	Analysis	Procedure	Program Documentation
Sales Replenishment		x	
Service Parts Control	x	x	
Stock or Not Stock		x	
Stock Catalog	x		
Stock Status		x	
Storeroom	x		
Tool Stores Inventory		x	

Production Planning

Decision Table applications in this area of activity include:

	Analysis	Procedure	Program Documentation
Alternate Evaluation	x		
Bills of Material Generation		x	
Bills of Material Maintenance	x		x
Continuous Process or Select Periods		x	
Critical Unit Scheduling	x		
Determination of Machine Centers		x	
Determination of Machine Operations		x	
Equipment Requirements	x		
Equipment Selection	x		
Job Scheduling		x	x

	Analysis	Procedure	Program Documentation
Labor Routings		x	
Level Reporting		x	
Load Data Experience		x	
Load Forecasting	x		
Machine Loading		x	
Machine Utilization		x	
Min/Max Order System Machine Loading		x	x
Performance Reporting		x	x
Product Process Routing Determination		x	
Production Control Specifications	x	x	
Production Order Writing System	x	x	x

	Analysis	Procedure	Program Documentation
Production Planning Subroutines to Various Types of Material and Stock Situations	x		x
Production Planning Reports		x	
Production Priorities Calculation		x	x
Production Scheduling (Training Aid)		x	
Routing Input Logic	x		
Schedule Lists			x

	Analysis	Procedure	Program Documentation
Scheduled Operation Simulation	x	x	x
Selection of Manufacturing Processes	x	x	
Shop Load		x	
Toll Message Rating			x
Trunk Line Assignment		x	
Valid Machine Numbers	x		
Work-in-Process Production-Control	x		

Marketing

Decision Table applications in this area of activity include:

	Analysis	Procedure	Program Documentation
Backlog, EDP			x
Circulation	x		
Coding of Orders		x	
Competitive Bidding	x		
Computer Preparation of Quotations	x		
Contract Terms		x	
Customer Information Recording and Retrieval	x		x
Customer Support—Spares	x		
Data Input Criteria	x		

	Analysis	Procedure	Program Documentation
Describe Warehouse Operation to Management		x	
Distribution Simulation	x	x	
Editing Orders	x		
Estimates (Screen Those To Bid On)		x	
Experience Comparison	x		
Follow-up Mailing to Prospects		x	
Instructions for Field Personnel	x		

	Analysis	Procedure	Program Documentation
Issuing Insurance Policies/Changes	x		
Mail List Application			x
Maintenance of Customer Records		x	
Marketing Information System		x	
Marketing Simulation		x	
Marketing Strategy	x		
Operations Research	x		
Order Cancellation Procedure		x	
Order Entry	x	x	x
Order Processing Manual		x	
Order Status System	x		
Order Volume Analysis			x
Pricing/Price Changes	x	x	
Processing of Customer Applications		x	
Processing of Service Requests		x	
Product Distribution			x

	Analysis	Procedure	Program Documentation
Product Identification		x	
Product Line Matrix	x	x	x
Product Sales Analysis (Statistics)	x	x	x
Publication Distribution by Computer		x	
Sales Data Base	x	x	
Sales Forecasting		x	
Sales L. P. Program	x	x	
Sales Order File Update		x	x
Sales Reporting		x	
Sales Training Manual		x	
Selecting the Proper Shipping Point		x	
Special Pricing Agreements	x	x	
Standard Product Coding		x	
Standard Warehouse Practices	x		
Subscription Information System	x		
Travel Forecast	x		

Purchasing

Decision Table applications in this area of activity include:

	Analysis	Procedure	Program Documentation
Analysis of Purchasing Procedure	x		
Choice of Purchasing Document		x	
Equipment Ordering	x	x	x
Ordering Material Practices	x		
Ordering Supplies		x	

	Analysis	Procedure	Program Documentation
Procurement		x	
Purchase Order Follow-up	x		
Purchase Order Revisions	x		
Receiving & Inspection System	x		
Selective Inspection of Receipts		x	
Steel Tolerance Requirements	x		

Quality

Decision Table applications in this area of activity include:

	Analysis	Procedure	Program Documentation
Failure Analysis		x	
Inspection of New Products	x		

Production

Decision Table applications in this area of activity include:

	Analysis	Procedure	Program Documentation
Data Collection	x		
Dept. Location of Production Equipment	x		
Die Standardization		x	
Equipment Time & Reliability			x
Exception Reporting		x	
Generation of Manufacturing Specifications		x	
Instructions for Salvage of Loose Ends		x	
Manpower Forecasting	x		

	Analysis	Procedure	Program Documentation
Manufacturing Practices		x	
Material Rejection	x		
Milling Order Entry	x		
Numerical Control Programming		x	x
Process Sheet Maintenance	x		
Tool & Dimension Use		x	

Engineering

Decision Table applications in this area of activity include:

	Analysis	Procedure	Program Documentation
Automated Product Design		x	
Engineering Analysis		x	
Engineering Changes	x		
Engineering Data File Maintenance	x		
Engineering Progress & Projection System	x		x
Engineering Specifications	x	x	
Fan Design Programs	x		
Part No. Assignment		x	
Project Approval & Assignment	x		
Selection of Research Projects			x
Technical Specification Writing	x		

Personnel

Decision Table applications in this area of activity include:

	Analysis	Procedure	Program Documentation
Attendance	x		
Benefit Plans		x	
Computer Audits of Personnel Transactions		x	
Data Systems Directives & Personnel Operating Instructions		x	x
Employee Benefit Claims		x	
Employee Classification		x	
Employment Processing		x	
Employee Skills Inventory	x		
Group Insurance on Line	x		x

	Analysis	Procedure	Program Documentation
Leave of Absence Authorization		x	
Nurse Utilization		x	
Personnel Information System	x		
Personnel Operating Manuals		x	
Schedule Counciling of Students		x	
Security	x		
Sick & Annual Leave Mechanics		x	
Student Admission Application		x	

Miscellaneous

Decision Table applications in this unclassified area of activity include:

	Analysis	Procedure	Program Documentation
Acquisition Studies		x	
Appendices to Procedure Manuals		x	
Approval Authority	x	x	x
Approval Procedures		x	
Card Punch Instruction Sheets		x	
Categorizing of Records	x	x	
Central Mail File	x		
Choosing the Best Procedure Layout	x		
Claims Payment Program		x	
Claims Verification		x	
Completing Forms		x	
Computer Feasibility Analysis	x		
Computer Index Run		x	x
Converting Codes to Description & Vice-versa		x	
Credit Card Operation		x	
Curriculum Planning	x	x	x
Daily Transactions Edit			x

	Analysis	Procedure	Program Documentation
Data Communication Study	x		
Data Display	x		
DP Program Analysis			x
Data Transmission	x		
Determine Option Installation		x	
Development of Clerical Systems	x		
Development of Office and DP Systems	x		
Distribution of Memos		x	
Document Flow of Reports	x		
EDP Systems Definition	x		
File Update Rules	x	x	x
Form Selection Guide		x	
Forms Control	x	x	
General Claims Reporting		x	
Hardware Selection	x		
Information Selection for Reports	x		
Instructions for Clerical Personnel	x	x	

	Analysis	Procedure	Program Documentation
Investment Analysis		x	
Length of Stay Reporting		x	
Library Circulation System	x	x	
Masterfile Update	x		
Medicare	x	x	
Operation Memos		x	
Operations Research	x	x	
Passenger Reservations	x		
Payroll Disk Pack and Reconstruction		x	
Policy Manual	x		
Problem Definition	x		
Property Maintenance Program			x
Proposed Capitalization Approval Routing		x	
Reports Control		x	
Scheduling Computer Activity	x		

	Analysis	Procedure	Program Documentation
Selection of Plant Locations	x		
Shareholder Proxy			x
Signature Authorization		x	
Sources & Distribution of Data	x		
Specific Form Needed & Data Required	x	x	
Sub-Station Load Analysis		x	
Systems Training	x		
Teller Training Manual		x	
Traffic Data Recorder		x	
Transportation—Car Record	x		
Type Report to Issue		x	
Validating Sequence Reporting	x	x	x
Vehicle Registration System	x	x	
Writing Program Logic	x		
Zip Code Conversion	x